BRING BACK THE KING

Also available in the Bloomsbury Sigma series:

Sex on Earth by Jules Howard
p53: The Gene that Cracked the Cancer Code by Sue Armstrong
Atoms Under the Floorboards by Chris Woodford
Spirals in Time by Helen Scales
Chilled by Tom Jackson
A is for Arsenic by Kathryn Harkup
Breaking the Chains of Gravity by Amy Shira Teitel
Suspicious Minds by Rob Brotherton
Herding Hemingway's Cats by Kat Arney
Electronic Dreams by Tom Lean
Sorting the Beef from the Bull by Richard Evershed and Nicola Temple
Death on Earth by Jules Howard
The Tyrannosaur Chronicles by David Hone
Soccermatics by David Sumpter
Big Data by Timandra Harkness
Goldilocks and the Water Bears by Louisa Preston
Science and the City by Laurie Winkless

BRING BACK THE KING

THE NEW SCIENCE OF DE-EXTINCTION

Helen Pilcher

Bloomsbury Sigma
An imprint of Bloomsbury Publishing Plc

50 Bedford Square
London
WC1B 3DP
UK

1385 Broadway
New York
NY 10018
USA

www.bloomsbury.com

BLOOMSBURY and the Diana logo are trademarks of Bloomsbury Publishing Plc

First published 2016

British Library Cataloguing-in-Publication Data
A catalogue record for this book is available from the British Library.

Library of Congress Cataloguing-in-Publication data has been applied for.

ISBN (hardback) 978-1-4729-1225-1
ISBN (trade paperback) 978-1-4729-1226-8
ISBN (ebook) 978-1-4729-1228-2

2 4 6 8 10 9 7 5 3

Typeset in Bembo Std by Deanta Global Publishing Services, Chennai, India
Printed and bound in the USA by Berryville Graphics Berryville, Virginia
Chapter illustrations by Matt Dawson

Bloomsbury Sigma, Book Eighteen

for Amy, Jess, Sam, Joe, Mum and
Higgs the Dog Particle
… to the moon and back …

… and for my Dad …
… who gave me my love of wild things.

Contents

Preface 8

Introduction: Bringin' It Back 11

Chapter 1: King of the Dinosaurs 33

Chapter 2: King of the Cavemen 65

Chapter 3: King of the Ice Age 95

Chapter 4: King of the Birds 125

Chapter 5: King of Down Under 151

Chapter 6: King of Rock 'n' Roll 179

Chapter 7: Blue Christmas 211

Chapter 8: I Just Can't Help Believing 233

Chapter 9: Now You See It … 251

P.S. 281

Key References: A Little Less Conversation, a Little More Reading 283

Acknowledgements 292

Index 296

Preface

When I was a kid, we used to go on family holidays to the Jurassic Coast, where dark grey cliffs cast ominous shadows on the shingle-smattered beach. It was chilly, wet and windy. My brother and I were forced to wear itchy, woolly hats, high-waisted flares and unflattering cagoules. We drank tepid chocolate from a flimsy Thermos and sat on slimy boulders munching biscuits. My parents called it 'character building' and 'cheaper than a package deal'. I called it 'borderline pneumonia'. The sun never shone on the holidays of my childhood, but there was always a chink in the clouds. There was always the possibility that one day we might stumble across the remains of some prehistoric behemoth. For hidden among the rocks at Charmouth in Dorset are the fossilised remains of creatures that swam, walked and flew 200 million years ago – pterosaurs, 'Nessie-like' plesiosaurs and an armoured dinosaur called *Scelidosaurus*. How much I longed to find them. How much I longed to meet them.

But holidays came and went, hopes raised and dashed. I never found a *Scelidosaurus* bone, or any other fossil for that matter. But I never gave up. I kept going back, and now enjoy subjecting my own three children to the same brand of seaside sadism. Endowed with an uncanny fossil-detecting sixth-sense, they are orders of magnitude more successful than I ever was. With their sharp eyes and wilful determination, they brave the elements to eat ice cream on the coldest of days, and find fossils by the bucket load. More satisfying than any shop-bought souvenir, these fossils are a constant source of joy and wonder. They're also free, and apart from the ones found in China, are not made in China. My kids have collected hundreds of fossils but somehow the enigma of this prehistoric world keeps pulling us back, looking for another fix, and prompting us to ask questions such as: 'What were these creatures really like?', 'Could a human beat a *T. rex* at arm wrestling?' and 'Can we bring them back to life?'

This is a book, not about arm wrestling, but about whether or not we really can bring extinct species back to life. It is the story of the scientists who are trying to make it happen. It's about their ingenuity and dogged persistence; their reassuringly thick skins in the face of sceptics and critics who say de-extinction either can't or shouldn't be done. It's also the story of the animals they seek to resurrect; extinct species that once graced the Earth but that had been presumed lost forever. This book is not designed to be an exhaustive review of current de-extinction projects. Instead, I have unashamedly chosen to feature the species and projects that interest me most. Apologies to the world's ugly animals and to plants – you don't get much of a look in. The book starts in the late Cretaceous Period, 65 million years ago, from whence controversial claims for the existence of ancient biomolecules in dinosaur bones have been made, and finishes in the future, where de-extinction could help to enhance biodiversity. *En route*, it detours via Siberia in the last Ice Age, Mauritius in the seventeenth century and Graceland in the 1970s.

In my time I have been a scientist, a stand-up comedian and a serious science journalist. I have a lifelong love of fossils and quirky animals, and am a dab-hand at growing cells in dishes and tinkering with their DNA. When I first read about de-extinction a few years back, I was upset; not because I thought science had spiralled out of control, but because I wondered whether, if I had pursued my scientific career, I could have had a pet dodo by now. It's because of these interests that I find the prospect of de-extinction absolutely spellbinding. I now find myself on the outside of the laboratory looking in, watching as brilliant scientists push the boundaries of human knowledge and redefine what is possible. De-extinction, I hope to persuade you, is not something to be feared or resisted. It's a force for good, not a tool of the dark side.

Oh, and did I mention that there is a chapter about de-extincting Elvis Presley? The quick-witted among you will all too readily point out that Elvis is not technically extinct, to which I say, 'technically' you are correct. But 'extinct'? 'Dead'? Does Elvis know the difference? Fortunately for us, humans are not extinct, but aren't you curious to find out whether the same technology being used to de-extinct the woolly mammoth could be used to stage Presley's greatest ever 'Comeback Special'? Just so you know, no one is seriously planning to clone the King of Rock 'n' Roll, so the Elvis chapter is my folly, but it makes for an interesting thought experiment, and begs the question, 'Are you clonesome tonight?'

INTRODUCTION

Bringin' It Back

'High on a hill was a lonely goat-oh,
Lay ee odl lay ee odl lay hee hoo.
Lived on a mountain, very remote-oh,
Lay ee odl lay ee odl-oo.

Spent every moment, happy and free-oh,
Lay ee odl lay ee odl lay hee hoo.
Then she got killed by a falling tree-oh,
Lay ee odl lay ee odl-oo.'

… so goes the song of the last ever bucardo, a wild mountain goat called Celia, who lived and died on the vertiginous cliffs of the Spanish Pyrenees. It was 6 January, 2000. Just as people were taking down their Christmas decorations and recovering from millennium-sized hangovers, the stocky bleater was doing what mountain goats do best: springing deftly from boulder to boulder without a care in the world. An adult female in the prime of her life, Celia was a fine-looking beast. She had large curvy horns, the obligatory goatee beard and that permanent look of surprise that comes from wearing your eyes on the side of your head. Bigger than the average goat, Celia weighed as much as a washing machine, but was much more agile. She spent her days darting among the lanky pine trees that cling perilously to the precipices, stopping only to nibble the occasional blade of grass or cast an imposing silhouette against the clear, cloudless sky. But when the tree fell in the forest, did Celia hear it? It seems unlikely, but if she did it was already too late. The large tree came crashing down without warning with the hapless Celia standing directly in its path. She never stood a chance. The weighty trunk crushed her skull and Celia bleated no more. It was a sad, sad day. The very last of her kind, the bucardo went extinct …

… and that should have been the end of it. Extinction is, after all, forever. Game over. The end of the road. Full stop … but for a group of European researchers who had other ideas.

Ten months earlier, José Folch from the Centre of Food Technology and Research of Aragón in Zaragoza, Spain and colleagues hatched a plan to future-proof the bucardo (*Capra pyrenaica pyrenaica*). Fully aware that Celia's days were numbered, they reasoned that if they could collect a tissue sample from her while she was still alive, they could use her cells to create a clone. In the inevitable event of Celia's death, genetic doppelgängers could be created and the bucardo brought back to life.

It was a bold plan. For the bucardo, it meant a possible second chance. But for the world at large, it meant something much, much bigger. If Folch's plans worked and the bucardo was reborn it would mark a defining moment in the history of the Earth; an end to the finality of extinction.

First the researchers had to capture the feisty female, so they set up traps on the cliffs where she lived then retreated to watch the scene unfold. It was only a matter of time before the inquisitive creature trotted into the big iron box and the trap door sprung shut behind her. Ready and waiting, the team scrambled up the mountain and peered inside, where they found an unharmed if slightly bamboozled Celia wondering what was going on and who had switched the lights off. Under general anaesthesia,* the scientists then removed two tiny samples of skin – one from Celia's left ear and one from her flank – and fitted her with a radio tracking collar so they could follow her movements after she was released. Then, when the animal came round and they were satisfied she was alert and well, they let her go.

Celia's biopsied cells, in the meantime, were making the journey of a lifetime. Stored carefully inside tiny tubes, the samples were being driven away from the precarious cliffs. I spoke to Spanish veterinarian and researcher Alberto Fernández-Arias,† who helped coordinate the escapade. He told me 'the cells were so precious that we couldn't take any chances with them. So we put separate samples in two different cars and had them driven to two different labs. That way if one car had an accident, some of the cells would still make it.' Fortunately, for the cells and the drivers, the journeys passed without incident. In the laboratories, biologists grew the bucardo's cells in petri dishes to boost their numbers, then froze them carefully in tiny protective vials so that they could be revived and used at a later date. The bucardo's fate lay squirrelled away in a vat of liquid nitrogen.

Then in 2002, two years after Celia lost her life to her bark-clad nemesis, researchers removed and thawed some of

*The bucardo, not the scientists.

†It was Fernández-Arias who gave the last bucardo her name. He named her Celia after his girlfriend, whom he has since married. Fittingly the pair now have 'kids' of their own.

the vials. They went on to perform the same sort of experiment that in 1996 had produced Dolly, the world's most famous sheep and the first mammal ever to be cloned from an adult cell. Celia's cells with Celia's DNA were injected into goat eggs that had been stripped of their own genetic material. After a brief electrical jolt, the reconstituted eggs then began to divide. One cell split into two. Two cells split into four, and on it went until, a few days later, bigger bundles of dividing cells could be seen floating around in the petri dish. The team had created living bucardo embryos, each one a clone of Celia.

For their growth to continue, the best embryos were then transferred into the waiting wombs of surrogate mother goats, who then stoically tried to carry the developing animals to term. Most of the pregnancies failed, but one plucky surrogate managed to carry her cargo until she was fit to burst. On 30 July 2003, Fernández-Arias and colleagues delivered Celia's clone. A natural birth, they all agreed, was out of the question. The surrogate wasn't too posh to push; she, and the little clone that she carried, were too *precious* to push; nothing could be left to chance, so the team decided to deliver the kid by Caesarean section. In a room full of researchers wearing surgical gowns and masks, there was a sharp intake of breath as Fernández-Arias helped to gently prise the little kid from the surrogate's belly. As he held her in his arms he could see how beautiful she was. The newborn had tousled, toffee-coloured fur, wide brown eyes and delightfully wobbly legs. Her vital signs were good and her heartbeat was strong. The little clone seemed perfect.

But then, all too quickly, things started to go wrong. 'I knew almost as soon as I held her that there was some kind of problem,' says Fernández-Arias. The little animal began struggling for breath and became increasingly distressed. Despite every effort to save her, the clone died just seven minutes after she was born. An autopsy later revealed that her lungs were grossly deformed. The poor kid never stood a chance.

The bucardo, so briefly back in the world, went extinct all over again, giving it the honour not just of being the first animal to be brought back from extinction, but the ignominy of being the first animal ever to go extinct twice.

It's a bittersweet story but one that marks the beginning of exciting times. To bring an animal back from extinction is a rousing, dogma-smashing whopper of an incredible thing. In the entire history of life on Earth, it has simply never happened before. So special is the event that it's even got its own new word. When an extinct species is brought back to life, we say that it has become 'de-extinct'. It's a clunky word that doesn't exactly roll off the tongue, but it does the job. People had considered alternatives such as 'not extinct any more' or 'undead' but the former was too cumbersome and the latter too 'zombie apocalypse', so the term 'de-extinction' stuck.

When the researchers finally went public with the story of the clone's creation, through a 2009 peer-reviewed paper in an academic journal, reactions were guarded. 'Cloned goat dies after attempt to bring animal back from extinction,' ran an online headline in the UK's *Independent* newspaper. Media reports seemed to focus on how close the team had come to its goal, rather than celebrate what was a remarkable scientific journey. Although the little bucardo's life was short, that she was born at all is an immense achievement that reflects the hard work, scientific excellence and extraordinary vision of the scientists who created her. Sure, the researchers were disappointed that she didn't live longer, but then cloning has never been an exact science, and cloning an extinct species … well, that had simply never been done before. 'What we did was a very important scientific step,' says Fernández-Arias, and the only thing standing between him and a healthy bucardo clone is funding and time. 'From a practical point of view, I think bringing back the bucardo is entirely possible,' he says.

A few years on and few people seem to have heard about Celia's clone, of the bucardo so briefly brought back from the dead. But the bucardo is not the only extinct animal to have been resurrected to date. In 2012, Australian researchers briefly de-extincted an amphibian they dubbed 'the Lazarus frog' (*Rheobatrachus silus*), an unusual creature with one hell of a party trick (to find out what it was, you'll have to read Chapter 5). And as time and technology come to pass, more animals will follow. The goal is that these de-extincted creatures will go on to live long and healthy lives, that they'll breed normally, live in the wild and be able to recreate entire populations of animals that once were lost. Extinct animals, some scientists would have us believe, could find a home in today's world once more.

Currently there are around half a dozen ongoing projects around the globe, all trying to de-extinct different animals. In Australia, work on the Lazarus frog continues. In the United States, scientists are working to bring back the passenger pigeon (*Ectopistes migratorius*), an athletic, rosy-breasted bird that once flocked in the billions, and the heath hen (*Tympanuchus cupido cupido*), a stumpy avian wallflower that once lived in the scrubby heathlands of North America. In the United Kingdom, scientists are considering whether or not to bring back the so-called 'penguin of the north', the Great Auk (*Pinguinus impennis*), while in South Africa, they're trying to revive the quagga (*Equus quagga quagga*), a bizarre stripy equine that can score you 17 in a game of Scrabble. In Europe, they're trying to recreate the forefather of modern cattle, the intimidatingly big-horned aurochs (*Bos primigenius*). Meanwhile, in South Korea, Japan and the United States, three separate teams of scientists are trying to de-extinct that most iconic of beasts, the woolly mammoth (*Mammuthus primigenius*).

Exactly how researchers decide to revive 'this' or 'that' species depends on the species in question, and what we have left of it. Some projects, where similar, closely related species are still alive, are focusing on fancy breeding. The TaurOs

Project in the Netherlands, for example, is cross-breeding different breeds of living cattle to create something that looks like an aurochs. Some, like the ones that aim to resurrect the bucardo and the Lazarus frog, are using cloning. Others, like the passenger pigeon project, involve some rather elegant genetics. But they all boil down to the same magic, raw ingredient. Three little letters. DNA.

Genetti Spaghetti

DNA, or deoxyribonucleic acid to give it its right and proper name, is a famously twisted molecule. On decoding its helical structure in 1953, Cambridge University biologists Francis Crick and James Watson did what all Englishmen do after a good day at work. They went to the pub, where Crick uttered arguably the finest piece of pub banter ever spoken. Not 'it's your round', or 'can I have nuts with that?; instead, Crick said that they had 'found the secret of life.' It was a beautifully simple and eloquent summation of a brilliant, game-changing piece of science. DNA *is* the secret of life because it contains the instructions needed to *create* life, and a full set of instructions can be found in almost every single cell in our bodies in a membrane-bound blob called the nucleus. This biological Haynes Manual is called the genome. Your genome is unique to you. You acquired it from your mother and father the minute their egg and sperm met, melded and began to divide. The genome of the dodo is the full set of instructions for making a dodo, while the genome of a *Triceratops* is the full set of instructions for making a *Triceratops*. You get the picture. Each species has its own, unique genetic code and that code helps determine the way each living thing develops and grows. Our genomes influence the way we look, the way we age, even sometimes the diseases to which we succumb. So, given its importance, it's disappointing to note that, out of all of the billions of species that have ever lived, the vast majority have either lacked the foresight or simply couldn't be bothered to decipher the instructions for making their own kind. Even endangered species, such as pandas and pangolins, who arguably should know better, have failed to take an interest; human beings remain the only species ever to have decoded their own genome.

Fortunately for the de-extinction cause, us humans are pretty good at retrieving DNA from other animals. We have selflessly decoded the genomes of many other species and what we have found is this: given the complexity it encodes, DNA is a pretty basic molecule, shaped like a long, tiny, twisted ladder. The rungs of the ladder are made from pairs of chemical groups called nucleotides. There are just four different nucleotides: cytosine (C), guanine (G), adenine (A) and thymine (T), which in this book I sometimes refer to as 'letters'. The letters always pair up in a particular way. C always goes with G, and A always goes with T. It sounds simple, but the average genome is many millions of 'rungs' long with differing amounts of Cs, Gs, As and Ts all arranged in different orders. The human genome, for example, which is both big and clever, contains a whopping 3.2 billion pairs of nucleotides.

The average 400g (14oz) tin of Alphabetti Spaghetti contains 230 letters,* all added, the manufacturers tell me, in such a way that a representative number of each of the 26 letters of the alphabet can be found inside every can. But suppose they made some tins containing only Cs, Gs, As and Ts and called it Genetti Spaghetti.† They'd have to produce 14 million tins to make just one copy of the human genome. If all of the tins were opened and the pasta shapes laid out in a straight line, they would reach all the way from London, England to Sydney, Australia … and then back again. The order that the letters are arranged in is called a 'sequence', and the act of reading them is known as 'sequencing'. Both are terms that I use in this book, so don't be frightened of them, just think pasta.

---

*I know this because I had to pay my children to count them.

†I asked them if they would and, disappointingly, got nowhere. Nor were they interested in my husband's idea for 'Baked Genes'.

Real DNA letters are, of course, far smaller and less filling than their sauce-covered alternative. Forget straddling the world; if you laid out all the nucleotides found inside a single one of your cells, this time the line would just about straddle a stream. Just as spaghetti has to be packed into cans to make it more manageable, so too the DNA inside our cells is bundled into discrete units called chromosomes, of which we have 23 pairs. Dotted along these chromosomes are sequences called genes. Genes are important because they carry the instructions for making proteins, and proteins are important because they are an essential part of living things. I'll mention genes again, but don't panic, think pasta.

For as long as DNA lasts in any meaningful quantity, we can reach back in time to reveal its secrets. That means that scientists can extract and study DNA not just from living or recently deceased animals but also from creatures that died a long time ago. Museums, we now realise, are a treasure trove, not just of stuffed, boss-eyed specimens but also of DNA that can be extracted from them and used for research. Pickled animals in jars and stuffed specimens on shelves can be mined for the genetic information they still contain. Even fossils, sometimes, can be fair game. Ancient DNA has successfully been retrieved from fossil bones, teeth and toenails. It's been prised from elderly eggshells, feathers and, rather joyfully, from fossil poo. Today, the study of ancient DNA is helping to shed light on everything from domestication of the dog to the sexual preferences of our distant ancestors.

For the most part, scientists who study ancient DNA do so not because they want to de-extinct animals, but because they want to understand how life evolves and changes. De-extinction, some argue, is a distraction, a media-driven obsession that has little place in a respectable ancient DNA laboratory. But times are changing. Respectable scientists *are* interested in de-extinction. Through interdisciplinary research, it's now possible to marry the secrets of ancient

DNA with cutting edge genetic technology; in effect, to cut and paste the ancient DNA of animals long gone into the modern DNA of creatures still with us. Journalists, myself included, may well seem to be obsessed with de-extinction, but for good reason. Who among us wouldn't be inspired by a living, breathing woolly mammoth, if such a spectacle were possible?

Jurassic Spark

Like many people, my first encounter with de-extinction was when I went to see *Jurassic Park* at the cinema. I remember sitting in the dark, nervously nibbling popcorn and being alarmed that door-opening dinosaurs might already exist in some scenic island hideaway. Suddenly the rain-soaked British summer holidays of my childhood seemed quite appealing – familiar and mundane but safe and velociraptor-free. 'That film' has a lot to answer for. If you're feeling uneasy with the idea of bringing extinct creatures back to life, then perhaps your preconceptions (which I hope to challenge later) are in no small part due to the disaster-fest that is *Jurassic Park*.

Based on Michael Crichton's novel of the same name, Steven Spielberg's film imagined a wildlife park chock-full of de-extinct dinosaurs. But these were clever, ethical beasts. They instinctively knew to eat the corrupt computer programmer and the sleazy lawyer, but spare the innocent children. When the power fails, the dinosaurs run amok, totalling jeeps, ambushing lavatories and generally causing havoc. It's a brilliant, hugely popular film that has spawned sequel after sequel and grossed hundreds of millions of dollars worldwide. But it doesn't exactly inspire confidence in the concept of de-extinction or, for that matter, in science itself.

In the film, computer programmer Dennis Nedry steals and sells dinosaur embryos to a corporate rival, while the park's creator, John Hammond, is on the one hand a twinkly-eyed grandfather, and on the other a money-grabbing fat-cat who does his science in secret, and who only shares his discoveries after his theme park is set up and fully stocked with merchandise. It's not a good analogy for the way science works. But scientists – good, honest respectable ones whose research is transparent, publicly accessible and regulated by the most stringent ethical codes – have been thinking about de-extinction since way before *Jurassic Park*.

In 1980, entomologist George Poinar from the University of California at Berkeley found himself marvelling at an unusual piece of Baltic amber. In the course of his research, studying insects and their parasites, Poinar had travelled the world collecting creatures whose lives were cut short when they became trapped in the sticky oozings of ancient trees. Over time the tree resin hardened into amber with the hapless animals imprisoned inside and then, amazingly, many millions of years later, the golden nuggets of prehistoric life were found, each one a gilded window into the past. In his time, Poinar has come across some remarkable specimens – worker ants carrying food, bees smothered in pollen and flies whose final throes were *in flagrante delicto* – all perfectly preserved inside their amber tombs. Or so it seemed. When Poinar looked at his amber-entombed insects close up, sliced into thin sections under a microscope, the insects' bodily tissues were always in poor shape. What looked good from the outside was a bitty, disappointing mess on the inside. But then he came across one fly that defied expectation. At first glance, the tiny gnat was nothing out of the ordinary. It hadn't been doing anything special, different or X-rated when it died. But when his colleague, microscopist Roberta Hess, looked at it under the microscope she saw a level of detail never seen before in an animal of this age. Remarkably, after 40 million years trapped in amber, some of the fly's cells were still intact. Inside them Hess could even make out tiny specialised structures that would have helped the living cells to function. There were energy-generating structures called mitochondria and protein-making factories called ribosomes.

But best of all there were nuclei, the DNA-containing control centres of the cells. Hess left a one-word note on Poinar's office door – 'Success!' Two years later, they published their findings in the journal *Science*.

It's worth pausing for a moment to consider the scientists' exuberance. To that point, no one had ever gazed inside the intact cells of a million-year-old creature before. The level of detail was exquisite, but it was the nuclei that really got everyone excited. Hess and Poinar's discovery raised a provocative question. If the fly cells still contained nuclei, perhaps scientists would be able to tease DNA from it, and from that learn about ancient genes, evolution and the history of life on Earth.

But some scientists were prepared to speculate even further. Poinar's paper caught the vigilant eye of a secretive cabal of scientists and clinicians in Bozeman, Montana, who went by the name of the Extinct DNA Study Group. Here was a clandestine thinktank that was not afraid to think outside the box. They asked Poinar to join and he quietly accepted.

In the second edition of their *Extinct DNA Newsletter*, the group's founder, John Tkach, outlined an intriguing thought experiment. What if, many millions of years ago, there had been a hungry mosquito that dined on a dinosaur then became trapped in amber, with its last supper still inside its stomach. If one could recover a dinosaur blood cell from inside that mosquito and then transplant it into an egg that had had its own DNA removed, perhaps it might be possible to grow dinosaur cells in culture, or maybe, just maybe, it might be possible to grow a dinosaur.

'It was a fairly outrageous idea,' says virologist Roger Avery from the Virginia-Maryland College of Veterinary Medicine, who was a member of the Extinct DNA Study Group back in the 1980s. 'None of us took it too seriously.' Although fragments of DNA had been found inside the cells of a 40,000-year-old woolly mammoth, the techniques needed to extract, analyse and replicate DNA were still primitive. Back in the eighties, most thought the recovery of ancient DNA from fossils and insects in amber to be a fool's errand. Scientists scoffed at the idea, so the members of the Extinct DNA Study Group kept themselves and their ideas very much to themselves.

The group never seriously planned to de-extinct a dinosaur, but its members did enjoy speculating what might be possible. Concerned that dinosaur DNA could become contaminated with genetic material from other sources, they even wrote to NASA to ask if a clean room might be available should someone ever wish to try the experiment. NASA's response, if they ever sent one, has become lost in the decades that followed, and the Extinct DNA Study Group dissolved as discreetly as it had been formed. But its members remain, to my knowledge, some of the first scientists ever to speculate on the possibilities of dinosaur de-extinction. And if their ideas about dinosaur resurrection sound a little familiar, then that's because they are …

Round about the time the Extinct DNA Study Group was dreaming of dinosaurs, George Poinar received a visitor to his Berkeley laboratory; a tall, gangly man who was interested in Poinar's amber-entombed gnat. 'He was very pleasant and asked questions about bringing back life forms in amber,' says Poinar, but when the visitor left, Poinar thought little more about him. It was only years later when he received a telephone call from someone at Universal Studios in Los Angeles telling him he had been acknowledged in the back of a book that was about to be made into a film that the penny finally dropped. The visitor, a young Michael Crichton, had been so convinced by the possibilities of Poinar's work that he used it as the scientific basis for his novel, *Jurassic Park*.

It's the End of the World as We Know It

Science has now advanced to the point where de-extinction is no longer a fantasy, it's a very real possibility, but who or what to choose? Where to start? More than 99 per cent of all the species that have *ever* lived on Earth are no longer with us. That makes an initial starting list of over four billion contenders. It's a genuine embarrassment of riches, but the reality is that species come and species go. Extinction is an integral part of the story of life on Earth.

Today, the *idea* of extinction is widely accepted, but that wasn't always the case. In the seventeenth century the Archbishop of Armagh, James Ussher, used the Bible to work out that the Earth was less than 6,000 years old. Specifically, he calculated that God had created the world on 23 October 4004 BC, a Sunday, when technically he should have been resting … or going to church. The Earth was too young, churchgoers argued, to have experienced extinction and God would never let the lifeforms he had so painstakingly created go extinct. That would be pointless … like baking a tray of biscuits then wasting them on your children.

But then fossils made people see the world in a different light. In 1796, French naturalist Georges Cuvier presented a paper to the Institut de France on the fossilised jaw of an elephant-like creature he would later describe as the mastodon. Critics argued that the bone had to belong to a living species; that somewhere on this Earth, these enormous pachyderms still roamed. But Cuvier had another, more controversial explanation. The bone, he said, belonged to a creature that no longer existed. It was one of many species on Earth that had gone extinct, wiped out by some huge, catastrophic event. The Earth, he argued, goes through periods of sudden change where many different species become wiped out at the same time, a concept we now know as 'mass extinction'.

By Victorian times, scientists had bought into the idea of extinction, but they thought Cuvier was a bit of a drama queen. Instead of the Frenchman's disaster-movie scenario, most favoured a more subtle brand of extinction, where species disappeared slowly, one at a time. There was, they argued, simply no evidence for mass extinction. But times have changed. We realise now that extinction is happening more or less constantly. New species are evolving all the time, and older ones become extinct if they are out-competed or fail to adapt to change. There is a normal 'background' rate of extinction, but there are also times when extinction rates spike and normal events become rudely interrupted by some enormous, life-snuffing cataclysm. In times of mass extinction, over half of all species become wiped out. Life diminuendos from a cacophonous roar to the tiniest of whimpers. Vast amounts of biodiversity are lost, in geologically trivial periods of time. Palaeontologist Michael Benton from the University of Bristol has likened it to a crazed, axe-wielding madman hacking away at the tree of life and pruning it to within an inch of its life.

We know from the fossil record that there have been at least five mass extinctions since complex life appeared half a billion years ago. The first, which wiped out around 80 per cent of all species some 450 million years ago, was caused by an intense ice age and a dramatic fall in sea level. The third and most devastating occurred around 250 million years ago, when global warming and one of the biggest volcanic eruptions ever caused 95 per cent of all species to die. The fifth and most famous happened 65 million years ago when an asteroid slammed into Mexico's Yucatán Peninsula, putting an end to the dinosaurs, and three-quarters of all species at the time. Extinctions – big ones – happen.

So we have a choice. We can accept that extinction is part of life, that it is inevitable, and that all species, in time, come to pass. That is, after all, the way it's been ever since life began. Or we can shake things up and do it differently. Extinction, Celia the bucardo clone has shown us, doesn't have to be forever. Species can be reborn. We don't have to take extinction sitting down any more. In fact, we don't have to take it at all.

Dropping Like Flies

In fact, de-extinction could be arriving at exactly the time when we need it the most. All around us today, species are disappearing quicker than a springbok in a speedboat. A 2014 study in the journal *Science* found that 322 species of land-living vertebrates have vanished in the last 500 years and that populations of the remaining species are experiencing a 25 per cent drop in numbers. Life's not looking peachy for the invertebrates either, with two-thirds of these spineless wonders showing a 45 per cent decline in abundance. Even cockroaches, fabled for their supposed ability to withstand a nuclear holocaust, are struggling. 'These are shockingly high figures,' says biodiversity researcher Ben Collen from University College London, who co-authored the study. And these are just the species we know about. When under-represented groups are added into the picture, it's estimated that the number of Earth's wild animals has halved over the last 40 years.

This isn't the normal background rate of extinction. It is something much bigger, much more deadly. Depending on who you listen to, we are either on the brink of or in the full throes of a sixth mass extinction. All around us species, including flies, are dropping like flies, and many more are likely to follow. And do you know what the worst of it is? It's all our fault. Mass extinctions have occurred before, but they've never been of our own making. Collen's research suggests that current extinction rates are around 1,000 times higher than during pre-human times, and in the future this figure could increase by another whole order of magnitude.

It all began in Africa around two million years ago. Before this time, the continent was full of enormous carnivores: huge 'bear otters', sabre-tooth cats and giant bear dogs. But then early humans arrived. According to palaeontologist Lars Werdelin from the Swedish Museum of Natural History in Stockholm, fossil evidence suggests that the disappearance of so many massive meat-eaters coincides with the time when our early ancestors, members of the group *Homo*, switched from being mainly vegetarian to eating more meat. Competition with humans for access to prey could be what drove the large carnivores to extinction.

It's a pattern that repeats itself again and again. Animals are living happily. Man turns up. Animals go extinct. In the Pleistocene, the geological epoch that lasted from 2.5 million to 12,000 years ago, the continents were rife with their own weird and wonderful megafauna. North America had giant sloths and giant condors. Eurasia had its cave bears, woolly rhinos and woolly mammoths, while Australia had a 2-tonne wombat, a marsupial lion and the biggest kangaroo ever known.

Then we arrive on the scene, and it's game over for the big mammals. It happened 60,000 years ago in Australia, 30,000 years ago in Europe and 10,000 years ago in the Americas. It's an idea that was popularised in the late twentieth century by US geoscientist Paul Martin, who dubbed the phenomenon the 'blitzkrieg model'. And it happened on islands, too. After their mainland relatives were wiped out around 10,000 years ago, isolated populations of mammoths clung on in the tiny Arctic islands of Wrangel and St Paul, but then humans turned up a few thousand years later and the last of the mammoths vanished. 'Timing is the biggest single piece of evidence we have for mankind's involvement in the extinction of the Pleistocene megafauna,' says Ross Barnett from the Natural History Museum in Copenhagen, Denmark. The fossil record tells it all.

In more recent history, there are man-made records to back up the devastation we have caused. In the seventeenth century, we sealed the fate of the dodo (*Raphus cucullatus*) – that big-bummed, bungling caricature of extinction – when we trashed its habitat and boiled its unpleasant tasting bones (see Chapter 4). A hundred years later, we saw off

Steller's sea cow (*Hydrodamalis gigas*), a 9m (29.5ft-) long marine mammal related to the dugong. This gentle giant was discovered in 1741 by German naturalist Georg Wilhelm Steller when his expedition to the Russian Far East became stranded on the remote Commander Islands. Starving sailors harpooned the shallow-grazing plant-eaters with billhooks then butchered them on the beach. According to Steller, the beef and blubber tasted like 'the best Holland butter', and when the stranded sailors were finally rescued, news of this taste sensation spread. Future Alaskan-bound expeditions stopped off to sample the sirenian's delights, until in 1768, just 27 years after it was first discovered, there were no more sea cows left to kill.

In the nineteenth century, the great auk, a large flightless bird that looked like a penguin that had had a nose job, was heartlessly hunted to death on the islands of the North Atlantic for its feathers, meat and oil. On Funk Island, which sounds like a great place for a music festival but isn't, great auks were not only plucked alive, they were boiled alive on fires made of … oily, flammable great auks. The last British great auk, caught on the archipelago of Scotland's St Kilda in 1844, was inexplicably branded a witch and beaten to death, while on the Icelandic isle of Eldey, where the final colony lived, the last two great auks were strangled and their bodies sold to collectors. That was the end of the great auk.

Bucardos lived carefree in the Pyrenees for thousands of years, but in the nineteenth century, hunters decided their curvaceous, spectacular horns would look better mounted on walls than on the elegant beasts themselves. Celia's kind were hunted relentlessly, so that by 1900, fewer than a hundred animals were left in the wild, and although the bucardo subsequently became protected, it was too little too late. In 1996, there were just three bucardos left alive, and in 1999, there was only Celia. And though we'd like to believe that we've learned from the mistakes of our forebears, the list continues to be added to today.

In Taiwan, logging spelled the end of the Formosan clouded leopard (*Neofelis nebulosa brachyura*), while in the Atlantic's Cape Verde islands, the natives' desire for skink oil and meat sealed the fate of the Cape Verde giant skink (*Macroscincus coctei*), both animals being declared extinct in 2013. A year earlier, international conservation icon Lonesome George, superstar of the Galápagos Islands and the last Pinta Island tortoise, ambled off this mortal coil at the ripe old age of a hundred plus. But the fate of his kind was sealed in the eighteenth and nineteenth centuries when passing whalers removed more than 100,000 of the giant tortoises from the archipelago. The original ready meal, live giant tortoises were stashed in the ships' holds where they could survive without food or water for a year or longer. Their bodies were a source of meat, while water stored in their neck bags and urine were used as drinking water.*

We may have started out big, wiping out much of the ancient megafauna, but today, through our actions, we're responsible for the extinction of so many species big and small. A 2015 study by ecologist Mark Urban at the University of Connecticut suggests that if global warming continues on its current trajectory, then one in six species on the planet could face extinction as a result. And biodiversity loss isn't just increasing with the changing climate; its rate is actually accelerating with every degree the world warms. That means that if we carry on the way we are, climate change is poised to accelerate extinctions around the world.

Inscribed on the information panel outside the former enclosure of Lonesome George at the Charles Darwin

*They really did drink tortoise wee, but not neat. They weren't savages. The beverage was diluted to make it more palatable.

Research Station in the Galápagos National Park are the following words:

> 'Whatever happens to this single animal, let him always remind us that the fate of all living things on Earth is in human hands.'

We live in the midst of a global wave of human-driven biodiversity loss. Conservative estimates suggest that there are currently between five and nine million animal species living on the planet, but that we lose between 30 and 150 of these every day. We think ourselves immune but our time too, inevitably, will come to pass. For as long as humans continue to acidify the oceans, pump carbon dioxide into the atmosphere and warm the planet, we place our world and the fate of its inhabitants in mortal danger.

What Now, What Next, Where To?

It's up to us to decide what happens next. Sure, we should protect and conserve what we have, but if extinction is part of life on Earth, then perhaps *de*-extinction can be part of the picture, too. In 2013, environmentalist Stewart Brand and genomics entrepreneur Ryan Phelan organised a hugely successful TEDx event on de-extinction. If you missed it at the time, it's well worth watching on YouTube. Under the watchful eye of the world's media, the husband and wife team brought together cell biologists, geneticists, conservationists and ethicists to discuss the possibilities, practicalities and potential pitfalls of de-extinction. Opinions were as varied as the extinct animals themselves, with some arguing passionately in favour of the technology, and others arguing vehemently against it. 'The idea of the TEDx event,' says Phelan, 'was to get the public involved, so that they can play a role in shaping the debate around de-extinction.' It's with a similar goal that I write this book.

De-extinction has the potential to profoundly alter the way we think about our world. It is fascinating because of the possibilities it raises. When I tell friends I am writing about the subject, they inevitably ask a variant of the same question: namely, can you bring a *T. rex* … or a mammoth … or a dodo … or a Neanderthal … or a _________ (*insert dead animal of your choice*) back to life. In the chapters that follow, I attempt to answer that question, and to ponder another. Just because we can bring an animal back from de-extinction, does it mean we should? So, put any preconceptions to one side and think … If you could bring back just one animal from the past, who or what would you choose? It can be anyone or anything, from the King of the Dinosaurs, *T. rex*, to the King of Rock 'n' Roll, Elvis Presley, and beyond. But be careful what you wish for …

CHAPTER ONE

King of the Dinosaurs

As bad days go, it was a real humdinger. When he woke that morning, around 65 million years ago, Stan had little idea that a rock the size of Mount Everest was hurtling towards the Earth at 70,000mph. A short while later when the fireball hit, it shook the Earth with the force of six billion Hiroshima bombs, triggering shock waves and an earthquake so big that tsunamis spread around the world.

Standing thousands of miles away, Stan felt the earth move, but not in a good way. Looking towards the Yucatán Peninsula, he would have seen an ominous cloud of dust and ash climb into the sky and blacken the sun. The temperature began to drop. Day turned to night. A bitter, dark winter spread across the globe.

Up to this point, Stan had had little to worry about. A 20-year-old *Tyrannosaurus rex*, he was 7 tonnes of ugly. Thick-skinned and as long as an articulated lorry, he was the top dog of North America. Stan and his family weren't on any menu – they *chose* the menu; apex predators in a world that boasted an all you can eat smorgasbord of delicious dinosaurs, all Stan's for the chomping. *T. rex* was literally the tyrant lizard king.

But when the asteroid hit, everything changed. Condemned to endless darkness, all the sun-loving plants and algae shrivelled and died. Those that depended on them soon followed. Plant-eating dinosaurs, mammals and other creatures wasted away. And then the carnivores, Stan and his kind, weakened and starving, succumbed to the inevitable. It was a slow, systematic, global extermination. Over the next few hundred thousand years, three-quarters of all plant and animal species on the Earth became extinct. The dinosaurs went extinct. It was the end of Stan and the end of an era. Literally.

The Chicxulub asteroid, as it came to be known much later, after humans evolved and invented language, polished off the dinosaur-laden Cretaceous era and catapulted its successor, the Tertiary, firmly into the spotlight. It was bad news for the dinosaurs, but good news for the tiny, scurrying mammals that seized the day and toughed it out. It was the opportunity our most distant ancestors needed to wedge their furry feet firmly in the door of life. The demise of the dinosaurs paved the way for good things: the evolution of humans, you, me and, ultimately, the birth of Elvis Presley. But let's stop for a moment and spare a thought for Stan. Before the asteroid hit, dinosaurs had ruled the planet for 135 million years. Humans, with our paltry 200,000-year back story, are but impudent whippersnappers. The same space rock that gave mammals their lucky break removed some of the biggest, most inspirational, most enigmatic creatures ever to walk the Earth.

Burning Love

I've been in love with *T. rex* for as long as I can remember. When I was a child, I wanted to *be* a *T. rex* when I grew up. It's only biology and a few hundred million years of evolution that's stopping me from living that dream. But the *T. rex* of my childhood was very different to the one we know now. Back in the seventies, *T. rex* was an oversized, green, scaly lizard with enormous scary gnashers. He (all dinosaurs were male in my mind) stood pretty much upright, like he had a broom handle up his bum, and he dragged his sorry tail across the ground. Cold-blooded, cold-hearted, brawny but brainless, he lounged around in swamps and did little apart from roar, kill things and rue the day his pathetically puny arms stopped growing. But none of that mattered. *T. rex* was the only dinosaur I was interested in. He was the only figure, from a toy box bursting with other dinosaurs, that ever saw the light of day, because, for me, he was the king. How much I longed to meet him.

But as I got older, things began to change. *T. rex* was still a celebrity, but like so many other stars from the seventies, his reputation became battered by a series of image-shattering revelations. As new fossils turned up, researchers began to suspect that *T. rex* probably wasn't stupid, slow or cold-blooded. He was, in fact, an intelligent, highly successful predator whose stance was more horizontal than it was vertical. Forget a pole up the bottom; this beast could have balanced a tray of shot-glasses on its back. But worse was to come. Tyrannosaur fossils turned up sporting not scales, but a fine coat of feathery fuzz, and no one could be sure if *T. rex* was green, blue, pink, purple or any other colour. It was the final nail in the coffin. The *T. rex* of my childhood was gone. In its place was an incongruously feathered assassin that had more in common with birds than with Godzilla.

The problem is that although we know a great deal about *T. rex* from the fossil record, there are just so many questions left unanswered. There really was, or is, a *T. rex* called Stan. Like many dinosaurs, he was named after the man who found him, amateur fossil hunter Stan Sacrison who, in 1987, noticed a pelvis sticking out of a South Dakota cliff face.* It took a team of experts more than 25,000 hours to excavate and prepare the pelvis and another 198 bones to yield one of the most complete *T. rex* specimens ever found. Stan, who can be visited at the Black Hills Institute of Geological Research in Hill City, South Dakota,† is 70 per cent complete. But although his bones reveal many things, including his age, size, how he moved and what he ate, there is just so much we don't know.

For starters, Stan might have hatched from a beautiful egg, but we just can't tell. No one has ever found a *T. rex* egg. He might have been a playful youngster or a socially awkward teenager. He might have wowed the ladies, been a lothario or a loner, a brilliant dad or an absent parent. We just don't know. We know next to nothing about how dinosaurs behaved or interacted with one another. We don't know what colour they were and we certainly don't know the answer to that most age-old of *T. rex* questions: what were its little arms for? The best we can do is make informed, educated guesses, based on the fossil evidence that we have. But it would be very different if we had a real, live dinosaur to study.

If we were to choose one dinosaur to bring back to life, it can't be some pint-sized 'never heard of it' Thingymasaurus

---

*A few years after Stan's discovery, Sacrison's twin brother Steve also discovered a *T. rex* that he named … Steve. In a similar vein, there are tyrannosaurs called Sue, Celeste, Bucky, Greg and Wankel – the latter after namesake Kathy Wankel, a Montana rancher.

†If you can't make it to Hill City, you can buy a life-sized replica of Stan from the Black Hills Institute for just $100,000 plus crate and packing fee. Allow six months for delivery. 'Takes an experienced crew less than an hour to assemble.'

discovered in the 1800s, known only from a toe bone. It has to be a well-known, larger than life showstopper. Who better than *T. rex*, a Hollywood blockbuster of a dinosaur; star of countless movies including *Jurassic Park*, *The Lost World* and *The Land that Time Forgot*? Who else has battled King Kong, Dr Who and Homer Simpson *and* toured with a toddler supergroup?* Famous across the globe, familiar to adults and children alike, *T. rex* has a pull like no other dinosaur.

So what if I told you that a modern living dinosaur is not fantasy. Making a dinosaur is not impossible. It's not a task for beginners, and it's not something you should try at home. It's a task that requires a lot of concentration and expertise. Think of the trickiest thing you've ever done – taking an exam, learning a foreign language, assembling a piece of flat pack furniture without swearing – then multiply it by the biggest number you can think of, add infinity and learn to unicycle. Bringing back a dinosaur is a bit like that, only more science-y. It would be difficult. There would be many practical, intellectual and ethical hurdles to consider, many failures along the way. But there are respectable scientists out there who think it can be done, who truly believe that they can make or 'engineer' a dinosaur … sort of. But just because it *can* be done, does it mean it should be?

What Could Possibly Go Wrong?

Ask someone if they'd like to see *T. rex* brought back to life, and you're unlikely to be met with indifference. Unless you ask a particularly sullen teenager on a particularly dismal day, you simply won't hear, 'you know what, I don't really mind,' or, 'WHATEVS'. It's a Marmite thing. You

---

*Dorothy the Dinosaur is a life-sized pirouetting *T. rex* puppet with a jaunty sunhat shading dead, lifeless eyes, who tours with toddler supergroup The Wiggles. She eats roses, not children.

either love or loathe the idea. Personally, I'm keen. I'd have a pet *T. rex* in a flash, as long as I could train it not to eat my kids or leave Cretaceous crap on the kitchen floor. But some people think this a terrible idea, and their reasoning tends to go like this:

> 'Are you crazy? If we bring back the dinosaurs they'll eat us all! Mankind will be doomed. Run! Run for your lives!'

It's perhaps not surprising that some people think like this. After all, in 1905, when palaeontologist Henry Fairfield Osborn first described and named *T. rex*, he billed it as the greatest hunter ever to have walked the Earth. Later that year, the *New York Times* bigged it up further, describing *T. rex* as 'the royal man-eater of the jungle', 'the absolute warlord of the earth' and 'the most formidable fighting animal of which there is any record whatever.' Clearly this was in a time pre-Naomi Campbell, but public opinion, it seems, can be slow to change. But would one really try to eat us?

Researchers can deduce what dinosaurs ate by studying their fossilised teeth. The teeth of vegetarian dinosaurs are different from those of meat-eaters and omnivores. Long-necked sauropods, for example, had peg-like gnashers for stripping branches, while *Triceratops* had broad, closely packed teeth that it used for grinding plants. *T. rex*, on the other hand, had chompers described variously as 'serrated steak knives' and 'lethal bananas', up to a foot long and set deep in the skull, perfect for tearing and slicing. Analyses of Stan's skull have revealed that when *T. rex* slammed its jaws shut it could generate a maximum bite force of around 12,800 pounds. That's equivalent to being body-slammed by 50 fat Elvises all at the same time. Clearly these are not jaws to be messed with.

So *T. rex* liked meat, but to find out what kind, researchers have followed other lines of evidence. *Tyrannosaurus* tooth marks have been found in *Triceratops* bones, and *Triceratops*-like bones have been found inside a tyrannosaur tummy *and* inside tyrannosaur faeces.* It all suggests that *T. rex* liked to eat other dinosaurs, so if a *T. rex* could take down a *Triceratops*, it'd probably think very little of dispatching you or me. *T. rex*, it's generally thought, was an opportunist. Like big cats and other top predators of today, this dinosaur was probably a hunter when necessary and a scavenger when the pickings were easy. Whatever he was, he wasn't one for social niceties. In 2010, Nick Longrich from Yale University and colleagues noticed deep gouges in several *T. rex* bones that could only have been made by another *T. rex* as it fed. *T. rex*, they surmised, was a cannibal. Now if Hannibal Lecter taught us anything, it's that Chianti goes well with liver and fava beans … and that cannibals should not be trusted, so a living, breathing *T. rex* should be given an extremely wide berth. Let's put it simply. *T. rex* was not a fussy eater. As a mother, I like that. As a potential entrée, I don't.

So in theory, *T. rex* could eat us. But *would* he? Not if we didn't let him. Although *T. rex* had a relatively large brain, his would be no match for our crinkly, clever cortices. By using science, we could devise ways to ensure we wouldn't get eaten.

The most obvious solution would be to keep him locked up. Larger than an elephant, an adult *T. rex* would need an enclosure of at least pachydermal proportions. In the wild, elephants have very large home ranges that can be hundreds of square kilometres in size. And while captive elephants

*Much to the delight of children around the globe, dinosaur number twos can also become fossilised. Coprolites, to give them their official title, are sometimes found close to dead dinosaurs or inside their back passage. They don't smell, they're not messy, but they're not easy to flush either. Some are huge. One *T. rex* poop, discovered in Saskatchewan, Canada in 1998, is over 30cm (1ft) long.

and zoo animals can survive in confines smaller than their wild ideal, this doesn't make for happy bunnies. If we're serious about bringing *T. rex* back, we shouldn't skimp on his enclosure. It should be as big and secure as possible, and kowtow to any rider that promotes good behaviour. Dinosaur expert Darren Naish from the University of Southampton estimates an adult *T. rex* would have eaten around 100kg (220lb) of meat twice a week. That's equivalent to 1,000 quarter pounder burgers per sitting, but Naish recommends chucking in a couple of cows from time to time and letting *T. rex* prepare his own burgers. Contrary to his meaty reputation, *T. rex* might also enjoy a smattering of fruit and vegetables. African wild cats eat fruit from time to time, so there's every reason to suspect that *T. rex*, another apex predator, might do the same. The enclosure would also need to be interesting, as bored zoo animals can become depressed and unwell. Ideally it would have a mix of shady woodland for snoozing, and more open terrain for chasing dinner. Extras could be chucked into the mix – balls for chasing, scratching posts or rocks for sharpening claws, and chew toys and bones for exercising gnashers. But the best boredom-busting addition would have to be another *T. rex*. 'If you have one dinosaur, you run the risk of creating a permanently lonely animal,' says Naish. This, he advises, could drive extreme behaviour. It might prompt the dinosaur to vocalise and call for others – annoying for the neighbours, but not deadly. But it could also trigger aggression and the urge to escape. An angry *T. rex* on the loose would be a serious problem. Away from his regular supply of cows, humans might start to look like a tasty option.

So suppose you were to find yourself being chased by a hungry *T. rex* – could you outrun one? The good news is that, yes, technically any person could outrun a *T. rex*, because technically *T. rex* couldn't actually run. It's a pedantic point, and one unlikely to bring much relief if a 6-tonne dinosaur was slavering at your heels, but, to classify as running, both of its legs would need to be temporarily off the ground at the same time. And that didn't happen. Biomechanics expert John Hutchinson from London's Royal Veterinary College has calculated that, because of its shape, an adult *T. rex* would have needed more than 80 per cent of its total mass in its hind leg muscles in order to become airborne. That's one hell of a drumstick and not remotely possible. So *T. rex* officially couldn't run.

It could, however, power walk. Like human speed walkers, *T. rex* had muscly buttocks and comparatively puny ankles. Palaeontologist Heinrich Mallison from the Museum of Natural History in Berlin generated a computer model that predicted that *T. rex* reached its top speed by taking short but very rapid strides. A sprinting *T. rex* might be truly terrifying, but a mincing dinosaur? I'm not scared.

That said, it could mince pretty quickly. Current estimates, based on anatomy, suggest that *T. rex* could have reached top speeds of around 40kph (25mph). That's twice as fast as most humans can run, so while Usain Bolt might stand a fighting chance over 100m, the rest of us mere mortals would be unlikely to make it to the finishing line alive.

Hutchinson recommends the following. Bigger animals find it harder to run uphill than smaller animals as their size and gravity work against them, so run for the hills and when you get there, run up them in a zigzag pattern. *T. rex*, with its bulk and sticky-out head and tail, was not quick on the turn. Hutchinson has modelled the dinosaur's inertia, the force it would take to resist turning, revealing that *T. rex* would have taken a full second to turn 45–90 degrees. It might not sound like much, but it could make the difference between your life and its lunch. Next, if you can, head for somewhere with big obstacles. A charging elephant will pursue a person through a lightly thicketed bush by trampling everything in its way. But it's unlikely to give chase through, say, a thickly wooded forest or an IKEA car park on a Bank Holiday Monday. Finally, if all else fails, try tripping the dinosaur up. When an animal falls and hits the ground, the force of the impact depends on its mass and its height. A 6-tonne *T. rex* with a belly 1.5m (5ft) off the floor would hit the ground with a deceleration of 6*g*, considerably more than you'd experience even on the world's tallest rollercoaster. Humans black out at 4–6*g*, but for *T. rex* the effects could be catastrophic. With puny arms too feeble to break its fall, the impact would shatter *T. rex*'s ribs and pummel his internal organs. An adult *T. rex* would very probably die if it was tripped up while mincing.

From Bone to Stone

So on to important stuff. How to bring back the King of the Dinosaurs? As I mentioned previously, to de-extinct an animal you need a source of that animal's DNA. But all we have for dinosaurs are their remains, cast in stone.

Fossils are the preserved remains or traces of animals, plants and other organisms from the dim and distant past. Researchers estimate that the number of species known through fossils is far less than one per cent of all the species that have ever lived, meaning that fossils are only formed in very exceptional circumstances. To become one, an animal has, rather obviously, to die. There are no living fossils.* The body then has to be covered up quickly, and stay that way. Getting eaten by scavengers or being dug up is bad. Falling into a bog, river or ocean and sinking down into the mud at the bottom is good. Sediment can then smother the remains, allowing minerals to seep into the body and begin to replace the various biological, organic molecules from which life is made. Over time and under pressure, the creature's remains are slowly turned to stone. Dogma has it that when

*Apart from The Rolling Stones

fossilisation is complete any organic trace of the animal is gone. Fossils, we are told, are made from inert, inorganic minerals. Generations of schoolchildren have grown up believing this. They still learn it today. So with no biological information to go on, you'd be forgiven for thinking that bringing back the King of the Dinosaurs is a non-starter. Chapter over. Put the kettle on.

But that's not always the case. In 1828, a lady by the name of Mary Anning went for a walk on the beach at Charmouth in Dorset. It's one of my favourite places. With every passing tide, new fossils are liberated from the relic-rich rocks and mudslides, and can be handpicked from the shoreline. Anning had an uncanny knack for finding incredible things. She found toothy marine reptiles, pterosaurs and strange shelled creatures. At a time when women did little but wear flouncy frocks and giggle coquettishly, she grabbed her hammer, hoicked up her skirt and headed down to the beach to search out these 'curios', which she then sold to learned gentlemen and posh Victorian tourists.* Although the exact age of her finds wasn't known at the time, Mary recognised these creatures as both ancient and extinct, and her finds went on to help shape some of the most influential theories of the day, including Darwin's theory of evolution by natural selection. Scientist, accidental feminist, a woman in a boorishly male world, Mary Anning was in short, a bit of a heroine.

So there she was, that day in 1828, walking along the shoreline, when she spotted an unassuming grey rock. To anyone else it wouldn't have looked like much. It was smooth and rounded, with no fissures or flecks of crystal. Exactly the sort of stone, experience told her, that could be hiding something valuable. So she split it with her hammer to

*It's said that the tongue twister 'she sells sea shells on the sea shore' was written about none other than Mary Anning.

reveal a small blob of dark, sooty stuff. When she scratched the feature with her finger, it left a smudge on her skin. This wasn't a creature, she realised, but something that had been made by one. Anning had stumbled across a puff of ancient squid ink. Around 200 million years ago, the warm Jurassic seas of Britain were chock-full of squid-like creatures. In life, the squid would have fired a cloud of ink into the water to bamboozle an approaching predator, but in death, it made a useful addition to Mary's stationery cupboard. She took it back to her workshop in the family home at Lyme Regis, where she added a little water and turned the purple powder into a fine, sepia-like ink, which she then used to draw pictures of the fossils that she found. What Anning realised, ahead of her time, was that the fossil ink she used to write with was essentially the same as that fired from the bolshy squid's backside many millions of years before. The substance hadn't changed. An organic molecule from the Jurassic seemed as fresh as the day it was made.

We know today that the ink is made of a molecule called melanin, the same pigment that gives skin and hair their colour. 'It's an incredibly recalcitrant molecule,' says palaeontologist Michael Benton from the University of Bristol, who studies the stuff. 'It was obvious to palaeontologists of the day that melanin could survive for very long time periods indeed.' It's an idea that modern methods have since confirmed. A few years ago, US scientists showed that the chemical make-up of Jurassic squid ink is chemically indistinguishable from that of modern cuttlefish. Anning wasn't writing with some chemically altered, geologically ravaged, fossilised version of squid ink. She was writing with the real deal. Melanin, an organic molecule produced by a living creature, can survive the passing of millennia undamaged. We can't, of course, make dinosaurs from melanin, but it raises the question: if melanin can survive the process of fossilisation, what else can? What about soft tissue? Proteins? DNA? Could they survive too?

It's a question that another Mary has been grappling with for most of her academic career.

'Little Round Red Things'

It all began the day that Mary Schweitzer started seeing spots. Seeing spots is rarely a good thing. It can be a sign that you're overtired, that your little one has succumbed to chicken pox or, worse still, that your teenager is careering pimple-faced into puberty. Their arrival generally spells trouble. So when Schweitzer started to see them, back in 1992, she felt nervous. She was staring down a microscope at a thin slice of fossilised dinosaur bone, but what she saw beggared belief. For there, in among the channels that had once housed blood vessels, were tiny red spots. Lots of them. With their claret-coloured centres that resembled nuclei, they looked for all the world like the red blood cells found in so many other creatures alive today.* But how could they be? The tissue slice belonged to a recently discovered *T. rex* called Wankel Rex, which had been dead for 67 million years. Fossils, as everybody knows, don't contain red blood cells, nuclei or DNA. Yet here they were: colourful, seemingly organic structures. There had to be a reasonable explanation. Perhaps, Schweitzer thought, the 'little round red things', or 'LRRTs' as she called them, were caused by some unknown chemical reaction, or some other as-yet unrecognised phenomena. Schweitzer, then a rookie palaeontologist just starting her PhD at Montana State University, was certainly in no rush to tell her boss, the renowned dinosaur expert and Curator of Palaeontology at the Museum of the Rockies, Jack Horner. 'I was scared to death of Jack,' she tells me. But word got around, and a short while later, Schweitzer found herself standing by as Horner took a look at the samples for himself. After a long

---

*Mammals don't have nuclei in their red blood cells, but reptiles, birds and all other vertebrates do.

and awkward silence, he looked up and asked Schweitzer what she thought they were. 'I told him I didn't know,' says Schweitzer, 'but that the objects were the same size, shape and colour as blood cells, and in the right place too.' 'OK,' Horner replied, 'prove to me that they're not!'

It was a challenge that became the focus of Schweitzer's PhD and was to shape her entire career. Schweitzer went on to show that the LRRTs contained molecules that are found in red blood cells, including haem, the small iron-containing compound that makes red blood cells red and enables the much bigger haemoglobin protein to ferry oxygen around the body. And when extracts of the 'spotty' fossil bone were injected into rats, the animals' immune systems made antibodies to bind to and destroy the foreign matter. The antibodies recognised not the mammalian or reptilian forms of the haemoglobin molecule but the avian version. Given that birds are descended from dinosaurs, the results suggested that dinosaur haemoglobin – or breakdown products from it – was still present in the ancient fossil. It put Schweitzer in a pickle. At this point, she couldn't prove categorically that the LRRTs were red blood cells, but she couldn't prove that they weren't either. All she could say for sure was that some sort of biological molecule could be found in dinosaur fossils. Melanin had company. Here, apparently, was another organic molecule – haem – that could persist through geological epochs of time.

Then along came Bob. Bob is a well-preserved *T. rex*, whose 68-million-year-old bones were spotted sticking out of a sandstone cliff in the Badlands of Montana by field chief crewmember Bob Harmon from the Museum of the Rockies. In 2003, round about the time Schweitzer was finishing her PhD, Bob's bones were finally prised from the cliff face after a back-breaking three-year excavation. But the site was so remote that the finds had to be airlifted back to base by helicopter. One of them, an enormous metre-long thighbone, was so big that it had to be sawn in half and carried out in pieces. But what was bad news for the fossil record turned out to be good news for Schweitzer. Back at base, Horner boxed up all the little bits of bone that splintered off and sent them to her for analysis.

It took little more than a glance for Schweitzer to realise she was onto something special. For there, lining the inside surface of the bone was a thin layer of very distinct, unusual tissue. It was fibrous, filled with blood vessel channels, and different in colour and texture to the cortical bone that makes up most of the skeleton. In fact, it looked exactly like a type of reproductive tissue that is found in modern-day birds when they are carrying a fertile egg. 'It was really obvious to me,' she says. 'I said to my assistant – it's a girl and it's pregnant!'

Bob was Roberta. He was a she! It was like something out of a Kinks song. It was also the first time anyone had been able to determine the sex of a dinosaur with any certainty. Medullary bone, as it is called, is a short-lived tissue that offers female birds a ready source of calcium for the developing eggshell. But it was the first time anyone had spotted this in a dinosaur.

The Acid Test

To Schweitzer, the next step was obvious. Bone is a composite made from minerals and the fibrous protein collagen. In medullary bone, the collagen fibres are laid down in a characteristically random fashion that can be easily seen when the mineral component of the bone is dissolved away with acid. It was an experiment Schweitzer had done many times with bird tissue, but never thought to try with a dinosaur fossil. After all, why would she? Common sense suggested that if fossils were, as the textbooks said, made from mineral alone, then there wouldn't be anything left after the acid had got to work. So she decided to drop a tiny piece of the *T. rex* femur into weak acid just to see what would happen, and cautioned her assistant, Jennifer Wittmeyer, not to let the experiment overrun, otherwise the entire sample would dissolve.

But next morning, when Wittmeyer went to retrieve the sample, something odd had happened. She was expecting to find a solid chunk of bone, but the structure she retrieved with her tweezers was anything but. With the minerals removed, what remained was a springy, fibrous clump of something that had to be organic. 'It looked like a lump of chewed bubble gum,' says Schweitzer. Here was evidence that soft, pliable tissue could survive fossilisation; that organic material, possibly collagen, was lurking inside a bone from the time when dinosaurs walked the Earth. 'But it just wasn't possible,' says Schweitzer. 'There were lots of people way smarter than me, who had been doing palaeontology a lot longer than me, who said, it's just not possible. I was terrified.' So she had her assistant re-run the experiment several times, always with the same result – the stretchy tissue that survived the acid bath looked like the remains of medullary tissue from modern bird bones treated the same way.

Perhaps, Schweitzer reasoned, the phenomenon was a quirk of medullary tissue, so she repeated the acid test with bits of Bob's cortical bone. This time, the results were even more startling. When the acid was rinsed away, tiny hollow tubes could be seen floating around under the microscope. They were bendy, branching and transparent, like blood vessels, and inside them, little round red things that looked, once again, like red blood cells. Mary was seeing spots again. But these weren't the only cell-like structures she saw. In between the fibres of bone matrix were the distinctive outlines of another cell type called osteocytes. These are produced and become trapped in the bone's internal scaffolding, from where they stretch out long tendril-like structures to connect with other cells. It's these tiny, star-like protrusions that Schweitzer saw when she looked down the microscope. Bob, it seemed, appeared to contain a whole new level of organic material never seen before in a fossil. So impressive was the level of preservation in Bob's Cretaceous bones that the structures Schweitzer saw when she looked down the microscope, could – if you didn't know better – easily have been misattributed to an animal that had died weeks rather than millions of years ago. She published her cautiously worded results in *Science* in 2005, concluding that the exceptional levels of preservation she had seen in her dinosaur specimens could extend to the cellular level and beyond.'

Beyond! That's the key word here. Schweitzer had shown that soft tissue and cell-like structures could survive in dinosaur bones for millions of years. But beyond that? What of the molecules that make up cells? Protein, DNA and the like? Could they also survive the fossilisation process?

Schweitzer decided to look for collagen, the most common animal protein, hints of which she suspected she had seen in the demineralised *T. rex* bone. Collagen is a structural protein, found in bone, muscle, skin and other tissues. Known for its durability, it was a sensible first choice to look for.

Knowing full well how controversial the discovery of dinosaur collagen would be, were she to find it, Schweitzer decided to chuck as many different tests at the *T. rex* bone as she could. Under high-powered microscopy, the substance bore stripes, each a precise 67 nanometres apart (a nanometre is one millionth of a millimetre), a pattern that is seen in and diagnostic of modern-day collagen. And when antibodies to chicken collagen were sluiced across the sample, they stuck. Because dinosaurs evolved from birds, it was thought that their collagen molecules would be similar. So the study hinted at the presence of something collagen-like in the dinosaur bone that could recognise and bind to the modern bird antibody.

And then the icing on the cake. With the help of colleague John Asara from Harvard's Beth Israel Deaconess

Medical Center, samples were run through a mass spectrometer, a seemingly magical black box of a machine that can be used to determine the constituent molecules that make up a protein. Seven different fragments of collagen were spotted. And by comparing the *T. rex* sequences to a database of modern animal proteins, the Cretaceous fragments were shown to be most similar to birds, followed by crocodiles – the two groups that are the closest living relatives of dinosaurs. It was the first molecular evidence for the long-held idea that birds are modern-day descendants of dinosaurs, and, more controversially, the first study to suggest dinosaur proteins could persist in fossils for tens of millions of years. 'It also,' speculated *The Guardian* newspaper, 'hints at the tantalising prospect that scientists may one day be able to emulate *Jurassic Park* by cloning a dinosaur.' Bringing back the dinosaurs just got one step closer, or so it seemed.

But the excitement was soon mired in controversy. Almost as soon as the results were published, scientists were lining up to knock them down. At the front of the queue, from the University of California, San Diego-based computational biologist Pavel Pevzner, who, in 2008, wrote what has to be one of the most butt-clenchingly scathing critiques ever penned in the name of science. He compared John Asara to a boy who watches a monkey banging away at a typewriter, eventually sees it produce a handful of words and then 'writes a paper called "My Monkey Can Spell!"' The alleged protein fragments were, Pevzner concluded, nothing more than statistical artefacts. Elsewhere, others suggested that Schweitzer had found bacterial scum rather than dinosaur proteins, or that the results were coloured by contamination from modern bird samples that Schweitzer had handled in her laboratory. Perhaps the anti-collagen antibodies were binding not to some ancient biomolecule, but to some inorganic alternative; an exquisitely detailed imprint or cast of the original protein moulded in mineral long ago, before decay set in. The coprolite hit the fan. Once again, just as when she started seeing spots a decade before, Schweitzer found herself having to defend her findings.

Help came in the form of a duck-billed dinosaur called Brachy. This 80-million-year-old plant-eating hadrosaur (*Brachylophosaurus canadensis*) had been recently discovered in Montana's Judith River Formation. Whereas Bob had been handled by umpteen sweaty, un-gloved people during its three-year excavation, and then coated with varnish to protect the bones, Brachy was given full CSI treatment. Concerned that human cells could contaminate the scene, that ancient biomolecules might decay quickly once out of the ground and that preservatives might permanently alter the fossil's chemical make-up, a protectively clothed Schweitzer and co. jackhammered the dinosaur from the surrounding rock in just three weeks, from whence its varnish-free remains were whisked to a custom-built mobile lab set up in the field. Once again, using multiple tests, signs of collagen were found – evidence, Schweitzer believes, that proteins or fragments of them really can, in very special circumstances, survive for 80 million years.

That was 2009. Today, Schweitzer's finds are still controversial. She still has her sceptics, but others are beginning to replicate her results. Signs of collagen and red blood cells have since been found in other fossils, and Schweitzer believes she has found other structural proteins hiding in her samples. Cumulatively the finds suggest that the preservation of organic matter in fossils may be more common than was originally thought. 'I believe we are coming to a point where people will need to acknowledge that this is the case,' says Sergio Bertazzo from Imperial College London, who was among those to find evidence of organic matter in dinosaur bones. The dogma that fossils are just made of rock looks set to crumble.

That scientists can retrieve organic molecules from fossils is impressive. Through their study, researchers hope to shed light on the biology and evolution of extinct creatures. But although dinosaurs were made of protein (and many other molecules besides), we can't somehow rebuild one from a few scrappy bits of collagen. It's like trying to construct the 5,195 piece Lego Star Wars *Millennium Falcon* (Model 10179) from just a few bricks and the picture on the box. Without the instructions it's impossible to know what the other bricks should be, or how to put them together. The rotating laser cannons could end up stuck on the detachable cockpit cover and Chewbacca could end up with Leia's hair. Unthinkable! So while a working knowledge of dinosaur proteins can help put us on the right track, we can't rebuild a dinosaur without the complete instruction manual. We still need dinosaur DNA.

The Hunt for Dinosaur DNA

The odds of finding dinosaur DNA are, however, stacked against us. The problem is that, unlike collagen, DNA is a hopelessly flimsy molecule, in some ways not unlike the Forth Bridge. According to the myth, Scotland's most famous rail bridge is so big that by the time workmen have finished re-painting it, it's time to start all over again. 'Painting the Forth Bridge', or, if you live in the US, 'Painting the Golden Gate Bridge', has become a colloquial expression for a task that never ends. So too our DNA must be continually repaired to keep it in good working order. Fortunately for us, the repairs are done automatically. As long as we're alive, our cells patch things up, repairing breaks in the DNA strand and replacing nucleotides that have become altered or lost. As we age, however, our cells struggle to keep on top of this routine maintenance, and the gradual build-up of errors that results is thought to contribute to ageing, and age-related diseases like Alzheimer's. When we die, the renovations stop altogether. As our cells and tissues decompose, so too does our DNA. Fragmented into ever smaller pieces, the genomes of dead things become more bitty and less recognisable until at some point in time, they are gone forever.

But that hasn't stopped people from looking. In 1984, Russell Higuchi and Allan Wilson from the University of California, Berkeley, became the first to retrieve DNA from a long-dead animal when they extracted the molecule from a bit of 140-year-old muscle from an extinct beast called a quagga. Then a year later, DNA was prised from a 4,000-year-old Egyptian mummy (see Chapter 2). It was the start of a research field that became known as 'ancient DNA'. Up until then 'ancient DNA researchers' were simply elderly geneticists. But now they were anyone, of any age, who researched elderly DNA. At this point, no one knew just how long DNA could survive. So the race was on to push the boundaries of DNA retrieval from ever-older specimens.

In the early 1990s, around the time that Schweitzer first started seeing spots, researchers claimed to have isolated DNA from a whole host of antediluvian remains. DNA was apparently retrieved from 17-million-year-old magnolia leaves, a 30-million-year-old termite and from a 120-million-year-old amber entombed weevil. Geological epochs fell like tripped up toddlers. But then came the icing on the cake. In 1994, scientists claimed to have retrieved DNA from an 80-million-year-old dinosaur bone.

It all seemed too good to be true. Such extraordinary claims left many academics scratching their heads. Surely it was impossible for DNA to survive that long? In the end it took Nobel Prize-winning biochemist Tomas Lindahl to point out that because of the way DNA breaks down, it simply can't survive over these kinds of time frames. And in 2012, his ideas were corroborated by a study that found that DNA has a half-life of just 521 years. Although it might sound like a long time, what this means is that after 521 years, half of the links between the DNA letters in a genome would be broken. Half a millennium later, half of those remaining links would have disappeared, and so on. After 6.8 million years, every single link would be destroyed, making the recovery of DNA from fossils any older than this completely impossible. The fantastical claims from the early nineties amounted to little more than wishful thinking mixed with an unwanted dash of contamination. There was no DNA in the fossils. Instead, the technique used at the time to amplify the fragmented bits of DNA, the Polymerase Chain Reaction (PCR), had accidentally amplified bits of contemporary DNA from the surrounding environment.

Twenty years after these wayward claims and researchers now have new techniques at their fingertips (see Chapter 2). They have honed their methods and can be confident that when they do find DNA in ancient samples, it's the real deal and not some annoying artefact. Lindahl's predictions have stood the test of time. At the time of writing, the record for the world's most ancient DNA goes to a 700,000-year-old horse found frozen in the Canadian permafrost. And the oldest human DNA to be salvaged comes from the femur of a 400,000-year-old hominin* found in an underground cave called the Sima de los Huesos, or 'Pit of Bones', in Spain's Atapuerca Mountains. We now have evidence to suggest that in exceptional circumstances (cold, dry places such as the Arctic or the inside of caves) fragments of DNA can sometimes persist for hundreds of thousands of years. 'If you're in an area where your ice cream will last, then your DNA will last too,' says ancient DNA researcher Tom Gilbert from the University of Copenhagen. But beyond that? Most ancient DNA researchers think the molecule

*A species of human that is more closely related to us than to chimps. Since we're the only human species left, all the other hominins are extinct.

can't possibly survive in fossils that are millions of years old. It's a brave person who sticks their neck out and still goes hunting for DNA in dinosaur bones.

But Schweitzer's not put off. 'If you can get DNA from a 700,000-year-old fossil, why not a million-year-old one?' she tells me, 'and if you can get DNA from a million-year-old fossil why not one that is seven or even 70 million years old?' I admire her chutzpah. Schweitzer has a hunch that could help explain how organic molecules can sometimes last for millions of years inside fossils. But better still, the same hunch could also help her to develop ways of flushing these molecules out.

Schweitzer thinks that iron is the key. Iron is a very reactive molecule, so in life it's kept locked up inside our cells. But after we die, our cells fall apart, liberating tiny iron nanoparticles that then generate highly reactive molecules called free radicals. Schweitzer thinks that the free radicals have preservative properties. They cause nearby proteins to become tangled up, making them more resistant to decay. 'In essence, they act a bit like formaldehyde,' she says. The process could help explain how it is that Schweitzer has managed to find signs of collagen in her dinosaur fossils. But it also made her wonder whether the same process could sometimes preserve other organic molecules too. What if the free radicals had preserved DNA? What if there really was DNA to be found inside certain fossils, but it had gone undetected because (1) no one believed it would be there so no one had bothered to look, and/or (2) it lay smothered under a layer of iron nanoparticles that effectively shielded it from view?

To test her ideas, Schweitzer used a chemical to remove the iron nanoparticles from her demineralised fossil samples, then ran two standard tests for DNA. Stains that fluoresce when they come into contact with DNA were sluiced across the samples, and when Schweitzer looked down the microscope, she could see tiny blobs of fluorescence inside what looked like the cells' nuclei. It looks just like the pattern of staining seen when the same test is done on modern cells, only fainter. The stains were clearly reacting with something. What's more, when she tested the samples with antibodies that bind to DNA, they too gave a positive result. The ancient cells lit up. Speaking to me more than a year down the line, Schweitzer remains rightly cautious. After all, no one else has ever found so much as a hint of genuine dinosaur DNA. Schweitzer's tests aren't designed to extract or decode the molecule. She's making no grand claims. Her tests are simply designed to see if bits of the molecule are still there at all. It's a sensible first step. Schweitzer's experiments reveal that there's certainly something inside her dinosaur cells, but all that can be said with certainty is that the substance is chemically and structurally similar to DNA. 'We've found a signature of DNA,' says Schweitzer, 'but we can't say for sure that it is DNA.' And although the substance was identified inside a dinosaur cell, she can't, with any certainty, say that the molecule is from a dinosaur. Ancient bone samples, as the researchers in the 1990s found to their cost, are all too easily contaminated with DNA from elsewhere.

But suppose the results are replicated. Suppose that sometimes, just sometimes, DNA does become preserved beyond the limits of what most think possible. The next step then would be to try to extract the molecule and de-code its chemical make-up; to 'read' or sequence it. But that is one very tall order.

Imagine printing out the lyrics of every song that Elvis Presley ever recorded – 700-plus nuggets of pop artistry – then putting them through a shredder multiple times. Hamster bedding! You sift through the detritus and millions of tiny scraps of paper fall through your fingers like confetti. You get fragments or suggestions of words or phrases, but no real way of reassembling them. The information is still there, but it's not in a form that can be used. The lyrics to 'Suspicious Minds' and lesser-known gems such as 'There's No Room to Rumba in a Sports Car', or

'Yoga Is as Yoga Does' are lost forever. In much the same way, Schweitzer may just possibly have detected the shredded remains of dinosaur DNA, but the *T. rex* genome, billions of letters long, cannot be reconstructed from these shattered remnants. The recipe for making a dinosaur cannot be found in fossils.

The Impossible Dream

Clearly, researchers need to adopt a different tack if they are to bring back the King of the Dinosaurs. Strangely enough, the answer may lie somewhere within the realms of *Jurassic Park* after all. Not in the film itself or in the methods it employed, but in the inspiration for the film's lead character. Schweitzer's boss, Jack Horner, is not only scientific advisor for the *Jurassic Park* films but also the inspiration for one of the movie's main characters, palaeontologist Alan Grant. Horner believes he could make a dinosaur within as little as a decade, without ever having to resort to ancient dinosaur DNA. 'All' he has to do is make evolution run backwards.

Imagine for a second that we could rewind the clock. Go back in time and similar, related creatures share common ancestors. The common ancestor of *Homo sapiens* and Neanderthals, for example, lived around half a million years ago. Go back a little further, however, and these common ancestors share common ancestors. That of humans and apes was around seven million years ago. Now keep on time-travelling; some 65 million years ago, we might meet the common ancestor of primates and rodents. And so it goes on; as we journey into the past, seemingly disparate groups of animals become united through their shared evolutionary history until we meet the last universal common ancestor of all life on Earth, a single-celled creature that lived 3.8 billion years ago. Run the clock backwards and you see evolution in reverse. The big, complicated tree of life becomes pruned into a much-simplified Brazilian landing strip.

Horner's argument is that if we take the modern-day descendants of the dinosaurs and then persuade evolution to run backwards, eventually we'll end up with a dinosaur. 'It sounds crazy,' says Horner, 'but it's not impossible.'

The first step, then, is to find a living descendant of the dinosaurs. That's the easy part. I have four of them at the bottom of my garden. They escape periodically, lay eggs when its suits them and harass my dog. But their tolerant nature does little to hint at their ferocious evolutionary past. Chickens – indeed all birds – are, as previously mentioned, descended from dinosaurs, specifically the two-legged predatory group of dinosaurs known as therapods, which includes both *T. rex* and *Velociraptor*. The split happened somewhere between 180 and 160 million years ago, when avian dinosaurs, or 'birds', as they became known, headed off down one branch of the evolutionary tree and therapods continued on another. The fossil record tells the story in exquisite detail. Early birds were, as you would expect, a halfway house between non-avian dinosaurs and modern birds. The most famous of them all, the raven-sized *Archaeopteryx*, had the wings and feathers of a bird but the teeth, snout, bony tail and clawed fingers of a dinosaur. Then, sculpted by natural selection and the passing of time, the teeth and tail disappeared, the snout turned into a beak and the wing took shape. While their forebears succumbed to the fallout from Chicxulub, modern birds took to the skies. The descendants of dinosaurs are all around us today.

The second step, therefore, is to take an embryo of a modern bird and persuade it to develop into something like its distant dinosaurian ancestor. It sounds far-fetched but for the fact that sometimes living, modern creatures display distinctly ancient characteristics. Emus are large, flightless birds with powerful legs and wings instead of hands. But peer at an emu embryo and you'll see that for a short time, it starts to develop the beginning of a three-digit forelimb rather than a single-digit wing. Elsewhere in the animal kingdom, snakes are sometimes born with hind legs, dolphins are sometimes born with hind fins and humans are sometimes born with tails. Known as atavisms, these features are throwbacks to the animals' shared evolutionary past. All are descended from the same four-limbed creature that hauled itself out of the water and onto the land some 390 million years ago. That these features sometimes exist at all suggests that the genetic instructions required to make them are still present. All Horner has to do is to find out what the instructions are and then find a way of reactivating them.

He outlined his ideas in his 2009 book *How to Build a Dinosaur*, and then again at a 2011 TED talk that has since racked up more than two million hits. By tinkering with the developmental programmes of embryonic chickens he hopes to persuade them to bring out their inner dinosaur; to develop dinosaur-like features like teeth and tails. His plans are founded on the realisation, formed over the last few decades, that although life exists in myriad different forms, all animals, from fruit flies to humans to chickens to dinosaurs, share similar genetic mechanisms that guide the way their bodies are laid out. The genes involved in making the compound eyes of fruit flies, for example, are very similar to those involved in the formation of say, your eyes or the soulful brown peepers of my faithful pet pooch. And most of the genes involved in making front paws in mice are similar to the genes that control the development of hands in humans. What's different are the patterns of gene activity; where and when, in the developing embryo, genes are switched on and for how long. This in turn is controlled by a complex array of control mechanisms, or 'genetic switches', if you like. The best-known of these are the *Hox* genes, a family of genes that code for proteins called transcription factors, which latch on to specific DNA sequences close to genes, so turning them on or off. Flip a genetic switch – alter the way that certain genes are expressed rather than the sequence of the genes themselves – and you can create animals that look very different to their natural selves, something Arkhat Abzhanov knows only too well.

Shape Shifters

Abzhanov isn't like Horner. He doesn't want to make a dinosaur. An evolutionary biologist at Harvard University, he's interested in finding out how evolution works, how tiny alterations at the molecular level craft much larger changes in anatomy. So he carefully snips a hole in the shell of an ordinary chicken egg, and places some tiny beads that he has pre-soaked in a special solution onto the face of the developing embryo. Then he watches and waits. A couple of weeks later, he sees something incredible, something not seen in birds for 90 million years. Instead of a beak, the tiny chick has grown a snout.

That was 2015. Now let's rewind a little to understand what went on. In 2004, Abzhanov showed that changes in gene expression can lead to subtle changes in shape. Tweak the levels of one particular gene in a chicken embryo and it develops a bruiser of a beak better suited to cracking nuts than it is to pecking grain. This led him to wonder if similar changes could trigger more dramatic anatomical changes, like the evolutionary transition from the snout of a dinosaur to the beak of a bird.

With no living dinosaurs available to study, he decided to plump for alligators, reasoning that the snouts of the two creatures would have developed in a similar way. The snouts of alligators, we know, are a set of paired bones called the premaxillaries. Birds have premaxillaries too, but during their development the bones fuse and elongate to form the beak. Abzhanov compared the embryos of both animals and found that two proteins, known to be involved in facial development, were expressed differently. In alligator embryos, patches of the molecules could be seen on either side of the face just before the snout began to form. But in chicken embryos, the molecules were spotted in the centre *and* the sides of the face. If the central patch of expression could be removed, he mused, then perhaps the chicken embryo would develop a snout. So he laced his tiny gel beads with molecules that bind to and inactivate the two proteins, cut a window in an eggshell and placed the beads right in the middle of the developing chick's face. Then he sealed the hole back up and popped the egg into an incubator. Sure enough, the chick developed paired premaxillary bones rather than a beak. 'It looks like the kind of snout that *Archaeopteryx* would have had,' says Abzhanov. So if chicken embryos can be coaxed to grow snouts, what else could they be persuaded to grow?

In 1999, Malcolm Logan and Clifford Tabin, then also both at Harvard, introduced a gene, *Pitx1*, which codes for a transcription factor, into the budding wings of normal chicken embryos. Instead of wings, the birds began to grow something a bit like legs. The odd-looking forelimbs contained leg muscles and had the beginnings of clawed fingers at their extremities. In 2005, when Matthew Harris, also at Harvard, altered levels of a protein called beta-catenin in embryonic chick jaws, he found that the birds grew teeth. They resembled the pearly whites of alligators, neat rows of conical teeth dotted along the little birds' jawlines. Here was a bird with bite, and proof that hens' teeth aren't as scarce as … well, hens' teeth.

Horner hopes to amalgamate all these findings and engineer these features – teeth, snout and forelimbs – into a developing chicken embryo. In addition, he has identified a number of genetic changes that he thinks caused primitive birds to lose their tails. Learn how to reverse these signals, he says, and one day we could persuade that same chicken embryo to grow a long, articulated tail instead of a parson's nose. Multiple changes, he acknowledges, would need to be made to multiple systems at different stages in development, but the end result, he hopes, would be a creature that looks like a coelurosaur, the subgroup of therapod dinosaurs that includes tyrannosaurs.

Horner calls his creation 'chickenosaurus'. It would be chicken-sized, feathery and, depending on how you look at it, either a pocket-sized dinosaur or one weird-looking Sunday roast. Horner hopes it would inspire a generation; a very modern dinosaur for a very modern age. Because gene expression will have changed rather than the genetic sequence itself, genetically speaking the creature will still be a chicken. According to Horner that means it should still behave like a chicken, posing no more danger than any of the other billions of domestic chickens alive today. And if it did get too feisty, we could always barbeque it ... so long as no one is expecting a wing.

I like Horner. I like the fact that where other palaeontologists deal in dust and bone, he's prepared to be different; to stick his neck out, to propose the audacious and to deal in DNA and flesh. But I can't help thinking that his ideas are a little naïve. While it's true that biologists have created dinosaur-esque features in chicken embryos, they're a far cry from the real deal. 'We ended up with a chicken with an odd looking hand,' says Malcolm Logan, now at King's College London, of his 1999 study. Because of ethical regulations, the embryos were never allowed to hatch, so there's no way of knowing if the limbs would have been functional or, indeed, if the embryos would have survived at all. What can be said with certainty is that the 'hands' that Logan engineered were a long way from anything authentically *T. rex*-like.

Also, while it's possible that the chicken genome does contain some sort of latent instructions for making snouts or forelimbs or tails, to presume that these are the same instructions that guided dinosaur development is a big leap. The fossil record reveals that many of these morphological transitions were already occurring some 230 million years ago as theropods evolved, meaning the genetic instructions for making them are not just ancient, they're positively Triassic. Chances are, because the instructions have lain dormant for so long, some of them will have been lost. Modern birds, we know, do not have the functional genetic sequences needed for making tooth enamel or dentine. But suppose, for the sake of argument, that the instructions are still there. What then? We realise now that genes don't act in isolation. Alter the activity of one and more often than not it changes the activity of others, sometimes with unpredictable results. We might persuade a chicken to grow a tail, only to find we have created unexpected problems. Other cells, tissues and ultimately body parts might fail to develop normally. The little creature might die inside the egg. Additionally, there's a huge amount of variation in how seemingly identical embryos respond to seemingly identical interventions, making Horner's ideas sound more than a little optimistic.

I'd love to see him prove me wrong but my belief is that, at best, Horner might be able to tinker with genetic pathways to create a crude caricature of a dinosaur. Ancestral-looking features might be recreated. The ancestral animal will not. What can be said is that the research underpinning Horner's dream is likely to yield benefits far beyond the realms of quirky-looking chickens. Unravelling the mechanisms controlling embryonic development is a fundamental goal of biology. Understanding how animals change from a single cell to a fully formed embryo can help us to understand what's happening when things go wrong. Research into the mechanisms controlling embryonic tail growth could lead to new treatments for spinal disorders. Research into the signals that guide limb formation could shed light on congenital limb abnormalities. That this type of research should continue is a no-brainer, but crowbarring it together to make an ugly chicken with teeth just doesn't do it for me. We can't and most likely will never be able to retrieve DNA from dinosaur fossils, so the authentic recipe for making a *T. rex* is probably lost forever. If I can't bring back a *T. rex*, I'll have to find something else to de-extinct.

CHAPTER TWO

King of the Cavemen

In Spring, poppies line the well-trodden footpath that leads the way to the Shanidar Cave. Scattered trees cling determinedly to the rocky hillside. Grasses smother the foothills. In the distance, looking down on those bold enough to make the climb is the gaping hollow of the cave's enormous, triangular mouth, carving a shadow deep into the mountainside. It's here, 60 years ago, that the bodies of seven adults and two children were unearthed. They were buried deep in the gloomy abyss and by the time they were found, tens of thousands of years after they fell, they were nothing more than bone. But by studying their remains, researchers have managed to piece together a remarkable story from long ago, one with all the hallmarks of a Hollywood blockbuster. It has violence and hardship, compassion and hope. It's a tale of immense courage and of survival against all the odds. But most of all it's a story that challenges our perceptions of what it means to be human.

The Shanidar Cave lies in the foothills of the Zagros Mountains in northern Iraq, and the bodies, painstakingly exhumed by US archaeologist Ralph Solecki and his team, belonged to an extinct species of ancient humans – the Neanderthals. This particular story concentrates on the first body to be found, an adult man whom they dubbed 'Shanidar #1', or 'Nandy' for short. His is one of the most complete Neanderthal skeletons ever found, but when they realised the state he was in, they baulked. The left side of his skull, which should have been smooth and rounded, was badly bashed in. His right arm, or what was left of it, was broken, withered and ended in a stump near the elbow. He had a broken bone in his foot, and arthritis in his lower right leg.

At first glance, the autopsy paints a picture of a lead character seriously down on his luck. The blow to his skull, maybe caused by a falling rock, would have left him struggling to see and paralysed down one side. The bottom of his right arm had been amputated, either deliberately or in some freak accident. His arthritis would have been a constant source of pain, and walking would have been difficult.

That we can tell so much about a life from bones alone is impressive, but there was more in store. The fact that Nandy's body had seen better days comes as little surprise – life in the Ice Age was, after all, no picnic. What's remarkable is that his broken bones show signs of having healed. Nandy recovered from his various injuries and went on to live for some time despite the difficulties with which they left him. Blind, paralysed and in constant pain, Nandy somehow managed to cope with his disabilities, eventually dying at a ripe old Neanderthal age of around 40. But there's no way he could have done this alone. After his injuries, Nandy would have been unable to hunt, run or defend himself. In a time of sabre-toothed cats and cave bears, this would have made him incredibly vulnerable. He would have needed help. Somebody (or somebodies) must have brought him food, kept him safe and helped to keep him going. In a time of need, Nandy had friends. Solecki concluded that Nandy was accepted and supported by his people up to the day he died.

After the brutality and suffering, the story has a happy ending. Nandy wasn't left to die. It's a little known fact that Neanderthals cared for their disabled. They looked after the elderly and nurtured the ill. The final credits go to the supporting cast of carers who helped our hero through his final years.

Meet the Family

Neanderthals are the undisputed King of the Cavemen. An extinct species of ancient human, they flourished in Europe and parts of Asia and the Middle East towards the end of the Pleistocene. Stocky, strong and ripped with muscles, they were similar to the East European shot-putters of today, but even hairier. In a time before razors and depilatory formulae, these prehistoric people hid their chinless faces under unkempt beards and their jutting brows beneath dishevelled dreadlocks. Sixty thousand years ago we met them in Asia, and after that in Europe. Then, around 40,000 years ago, Neanderthals disappeared (no one's exactly sure why), while us 'modern humans' trickled across the continents and slowly took over the world. Extinct they may be, but Neanderthals remain our closest human relatives. To bring back a Neanderthal would be to meet an old member of our family, to step back in time – and take an unprecedented sneaky peek into the story of human evolution.

It would certainly be intriguing to meet a Neanderthal face to face, to find out firsthand the similarities and differences between us. If he or she were clean-shaven and smartly dressed, would we even notice them if we sat next to them on a park bench? If we did, would the conversation flow? Could we become friends? Go out clubbing? Could a Neanderthal hold down a job, a marriage or a modern meal? How 'human' would he or she be? Whether or not we 'should' do it is a question I'll return to later, but for now, as thought experiments go, you have to admit it's a good one.

We've been fascinated by Neanderthals ever since they were first discovered, in a limestone quarry in Germany's Neander Valley, way back in 1856. The Victorians, who believed that God had created man in his image just a few thousand years before, struggled to accept that the primitive-looking skeleton, with its protruding brow ridges and thick bowed legs, could ever be human. Instead it was dubbed some kind of primeval 'missing link', a stupid ape-like creature with no redeeming features. Its barrel-shaped chest could never have squeezed into a whalebone corset. Ugly and unfamiliar, it could never had played croquet or taken tea. Neanderthals were, the Victorians concluded, savage and subhuman.

It was the beginning of a smear campaign that has plagued the Neanderthals ever since, not to mention one of the longest and most heated debates in modern science – how sophisticated or otherwise were these people? Even today, after archaeologists have pored over the remains of hundreds of different individuals and found copious evidence to the contrary, Neanderthals are still perceived as club-wielding, loin cloth-wearing savages. In popular culture, they're often displayed as unkempt cavemen who dragged their knuckles on the floor and communicated like modern-day teenagers – with the same inarticulate grunts and gestures. Even in conversation, the term is used in the pejorative. To call someone a Neanderthal is to call them a hirsute, stupid thug. They are the perennial butt of evolutionary jokes – a Neanderthal walks into a bar and says ... nothing. He can't even talk. Worst of all, they're not even here to defend themselves. Talk about kicking a man when he's extinct.

The fact is that from fossil remains and other lines of evidence we know more about this particular extinct hominin species than any other, and the Neanderthal stereotype is unjust, unfounded and unnecessary. Nandy's story shows us that Neanderthals were capable of compassion and camaraderie. An ancient human that cared for its sick, its disabled and its elderly is not so different from us. Far from savage, these people must have had deep social bonds, just as we do today. They looked out for one another, and cared for their comrades when in need. Perhaps they would fit in with us after all.

But could we ever bring back the King of the Cavemen? I should perhaps stress at this point that no one, to my knowledge, is seriously planning to resurrect a Neanderthal. But while researching this book, I have spoken to various highly qualified, well-respected researchers, all of whom tell me that making a Neanderthal is entirely possible – not with some as-yet undetermined technology, but with the know-how, kit and caboodle that we have today.

If nothing else, de-extincting Neanderthals might go some way to resolving the smear campaign against them, providing Nandy and his kin with some much needed positive spin. It would certainly help to inform the 'what were they like?' debate. While bone can survive in the fossil record, soft tissues and organs more often than not rot away. Resurrecting a Neanderthal would quite literally put flesh on the bones. We'd find out not just what they look like, but how they behave and what they can do. A project of this kind could have even grander goals. Harvard biologist George Church, of whom much more later, argues in his book *Regenesis* that resurrecting a fellow human species would do more than simply satisfy academic curiosity. It might, he argues, help us think about ourselves differently, and lend insights into other forms of human intelligence and different ways of thinking. There could be health benefits. What if Neanderthals turned out to be naturally resistant to, say, the HIV virus, or to the bacterium that causes tuberculosis? Insights gleaned from their biology could be used to help develop new therapies for our own species. By studying their embryonic development, we could learn more about the processes that guide our own. And if a Neanderthal ever did walk into a bar, I'd like to be there to buy them a drink, on the rocks, of course, and hear what they really have got to say.

A Terrible Smell

If cleanliness is next to Godliness, then Svante Pääbo is right up there with the cherubs on the clouds. The Swedish biologist has founded his career on a level of cleanliness bordering on paranoia. Pääbo is Director at the Max Planck Institute for Evolutionary Anthropology, a curvaceous, custom-built research facility in Leipzig, Germany, where he goes to extreme lengths to recover DNA from fossils. He's fanatical about keeping his samples, researchers and laboratory absolutely spotless in order to minimise the chances of contaminants tainting his experiments. As a result, he has changed the way we think about human history. He has prised DNA from a finger bone found in a Siberian cave to reveal the existence of a previously unknown species of human called the Denisovans. And he has sequenced the entire genome of the Neanderthal, revealing an unexpectedly spicy secret from our dim and distant past. Through his scientific rigour and attention to detail, Pääbo has arguably done more than anyone else to advance the field of ancient DNA.

Pääbo's career in ancient DNA took off after he cooked up a very unappetising piece of liver. It was the late seventies and Pääbo, who was working in virology at the time, had heard about some new techniques that enabled researchers to extract and sequence DNA. Obsessed with Ancient Egypt since his childhood, Pääbo found himself wondering if it might be possible to recover DNA from Egyptian mummies. At the time, the received wisdom was that the task was impossible, that the bandaged bodies would have been hanging around for far too long to have any DNA left in them. But Pääbo reasoned that if the enzymes that break down DNA need water to work then the molecule might still persist in ancient tissues that had been desiccated or mummified. He devised an experiment to test his idea, but because he was meant to be working on other things he decided to keep it a secret from his boss.

With no access to any Egyptian mummies, Pääbo had to make his own mummified tissue. So he went to his local supermarket, bought some cow's liver, then bunged it in the lab oven on a low heat and left it to fester. Twenty-four hours later the smell of putrefied flesh filled the room and his clandestine bake-off was in danger of being busted. But the air soon cleared and all that was left a few days later was a hunk of meat so dry that a Michelin inspector would have been more likely to check it for tread than award it a star. The meat may have been well past its sell-by date, but much to Pääbo's delight, its DNA was not. Desiccation had not destroyed all of the cells' genetic material. Small fragments could still be found.

If DNA could be found in over-cooked dinner, perhaps, he thought, it really could be found in the desiccated bodies of pharaohs. He was granted access to 20 mummies at the State Museum in Germany's East Berlin, but when he took samples and tested them for DNA, they all drew a blank … apart from one. Pääbo hit the jackpot with a 2,400-year-old mummy of a child. It was time, he decided, to come clean about his secret experiments. Pääbo told his supervisor and, with his full blessing, went on to publish his findings in the journal *Nature*, bagging the front cover with a mock-up of a mummy tastefully draped in DNA.

'The *Nature* paper created a real buzz,' says ancient DNA researcher Tom Gilbert from the University of Copenhagen in Denmark. 'People were really excited.' With Pääbo's finding, the prospect of DNA from the deep past was no longer a pipe dream. Suddenly, everyone wanted in on the action. There followed what has since become known as the Wild West years of ancient DNA research, where it seemed anyone with a passing fancy in ancient DNA chanced their arm at extracting it. Claims of 'antediluvian DNA' hit the headlines (see Chapter 1). Researchers claimed to have found DNA from fossil plants, insects and dinosaurs, but it all came crashing down when researchers realised that modern DNA, most likely from the people who had handled the samples, had tainted the results. Extreme measures were required. If Pääbo were to trust his results, he had to be sure that the DNA he was finding was genuinely ancient and not some modern contaminant. So it's here that his penchant for the pristine began in earnest.

At the University of Munich, he built the world's first 'clean room' for ancient DNA, a small windowless cell that he scrubbed with bleach and kept squeaky clean. Anyone working there had to wear CSI-style overalls, gloves, face masks and hairnets in order to make sure that none of their own DNA contaminated the samples they were studying. To be on the safe side, Pääbo decided to hone his techniques on extinct animals, rather than ancient humans, reasoning that even if he did find human DNA in a fossil, it would be impossible to tell if it came from an ancient person or a member of his own lab. So when he successfully isolated snippets of DNA from a 25,000-year-old horse, and then from a 50,000-year-old mammoth, he was pretty certain that the DNA he'd found was genuine. There were, after all, no Ice Age horses or mammoths roaming around the lab.

It was only then that he turned his attention to Neanderthals. They seemed like a good choice to study for many reasons. For a start, they are our closest evolutionary relatives. They lived recently enough that their bones could still contain DNA, yet were still reassuringly ancient. Studying the genetic differences between Neanderthals and modern humans would, he hoped, enable researchers to identify the key changes that set our earliest ancestors aside not just from Neanderthals, but from everyone and everything else on the planet. It could help explain why Neanderthals produced rudimentary tools and hunted mammoths, while we acquired the capacity to text and dial a pizza. It could help clarify how Neanderthals were capable of culture, art and abstract thought, yet never created social media or the Sistine Chapel. The differences could help explain why Neanderthals disappeared while we flourished. From Neanderthal DNA, Pääbo hoped to glimpse the biological origins of being human.

Pääbo was offered access to a German national treasure, the first ever Neanderthal to be found, which was kept at the Bonn Museum. Using a tiny sample from the fossil's upper right arm, Pääbo was able to work his magic again. Sixteen years after he had extracted DNA from his shop-bought liver, he became the first scientist to isolate DNA from an ancient human.

Like all of the reputable ancient DNA studies that had gone before, Pääbo chose to extract DNA not from the cell's nucleus but from its other hiding place, tiny energy-producing structures called mitochondria, where DNA is more plentiful. A single cell can have thousands of mitochondria, and a single mitochondrion can have hundreds of copies of DNA. Compare that with the two DNA copies found inside the one nucleus of a regular cell, and mitochondrial DNA seems like a comparatively low-hanging fruit. So it was mitochondrial rather than nuclear DNA that Pääbo had extracted from the Neander Valley specimen. It was impressive, but if researchers were really to unravel the genetic secrets of the Neanderthal, they would need more than just mitochondrial DNA. Genomes, we know, are made from mitochondrial DNA *and* nuclear DNA, but it's the nuclear DNA that makes up 99.9995 per cent of the total sequence. So if Pääbo were to decode the Neanderthal's genome in its entirety, he would need to be able to retrieve DNA from Neanderthal cell nuclei.

But at this point in time, no one knew if that was possible. In the end, it took some remarkable poo to settle the matter once and for all.

Top of the Plops

In the late nineties, Hendrik Poinar, a Californian geneticist working in Pääbo's lab, was studying 'molecular coproscopy', or, to give it its proper name, 'the study of fossil poo'. Top of the plops for Poinar were the ancient faeces of the extinct Shasta ground sloth (*Nothrotheriops shastensis*), a bear-sized vegetarian from the last Ice Age. Like so many of my previous boyfriends, it had long, coarse hair, prehensile lips and walked on its knuckles. Unlike my exes, however, it ate a lot of plants and left a lot of droppings in the dry caves of what is now the southwestern United States. At one site, the Rampart Cave in the Grand Canyon, Arizona, the floor is literally wall to wall dung balls. Now faeces may not be a living creature, but they still contain DNA, and over millennia on the cave floor the sloths' dung had become fossilised.

Poinar studied the chemical composition of the poo and found that it bore the tell-tale signs of a reaction more frequently found in the kitchen. The Maillard reaction, as it is known, occurs when common sugars are warmed for a long time. Sugar molecules form chemical cross-links with proteins and DNA, causing large, tangled complexes that give browned foods, such as bread and steak, their flavour. Poinar's experiments suggested that many millennia ago, as the warm dung lay slowly fermenting on the cave floor, the same reaction had occurred. Ancient DNA had become tangled up with other molecules, meaning that the method used to detect and amplify it – the Polymerase Chain Reaction (PCR) – simply wouldn't work. But the Maillard reaction can be reversed. When a particular chemical is added to baked bread, it breaks the tangles apart and returns the loaf to dough. When Poinar added the same chemical to extracts of his coprolites he found, to his delight, that the fossil's inner DNA could be liberated. He found not just mitochondrial DNA but nuclear DNA, too. The experiments suggested that sometimes nuclear DNA could survive inside and be extracted from ancient fossils. And if nuclear DNA could be retrieved from Ice Age poo, perhaps it could also be retrieved from other Ice Age denizens, including Neanderthals.

Ultimately, it would take a few years and some serious new kit to turn this dream into reality. One big problem was the method used to retrieve DNA from the ancient samples. To that point, researchers had used PCR, which amplifies tiny fragments of DNA into larger amounts so that it can be studied. But the procedure is notoriously prone to contamination. If there is even the tiniest bit of modern DNA in a sample – from a skin cell, say, belonging to the researcher who handled the samples, or from a bit of dust in the laboratory – then very often that is the DNA that gets amplified. Modern, contaminating DNA can swamp a sample and wreck it as a result. A tedious and untrustworthy process, Pääbo realised that new techniques were needed.

They arrived in the noughties in the guise of 'Next Generation Sequencing (NGS)'. It sounded very *Star Trek*, and just like the TV show quickly built a loyal fanbase of geeky types who preferred it to the original. It was sequencing, Jim, but not as we know it. Where PCR focused on just one specific DNA fragment at a time, NGS could read, or 'sequence', all of the many thousands of tiny DNA fragments in a sample in one go. It did away with the troublesome PCR step, so researchers didn't have to worry about modern contamination drowning out the DNA they were interested in.

Pääbo was so impressed, he bought his own NGS machine, and after sequencing a respectable million base pairs of Neanderthal DNA, announced to the world that next he would sequence the entire nuclear genome. But there was another problem. Modern contamination might not have been an issue anymore, but *ancient* contamination was. Pääbo realised that the Neanderthal samples he cared about so deeply tended to contain around 3 per cent Neanderthal DNA and 97 per cent DNA from bacteria that had, at some point, feasted on the bone. With NGS sequencing *all* the DNA in a sample, it meant that most of the data generated would be of no interest at all.

In the end, Pääbo had to add two new steps to his methods: one to physically remove the unwanted ancient DNA, and the other to amplify the little Neanderthal DNA there was. With his new improved protocol the Neanderthal genome was well on its way, but Pääbo and his team faced one last challenge. Where modern DNA is long and stringy, ancient DNA is broken and bitty. Most fragments prised from millennia-old fossils are just tens of nucleotides long. How, then, to reassemble these tiny pieces into a genome three billion letters long?

Imagine the largest, most complicated jigsaw that you've ever done. Then imagine that many of the pieces are damaged, and many others are missing. Annoyingly, there are also extra bits that don't seem to fit anywhere, and random additional copies of pieces that do. Reassembling the Neanderthal genome is a bit like doing that puzzle. According to Amazon.com, the world's most difficult jigsaw puzzle is 'The Sweet Shop', a double-sider featuring Smarties. The Neanderthal genome, in contrast, is a three-dimensional puzzle featuring nucleotides. And where 'The Sweet Shop' has 529 pieces, the Neanderthal genome has hundreds of thousands.

Just as the Amazon puzzle would be tricky to fit together without the picture on the box, so too Pääbo and his team needed some sort of reference to help them arrange the bits of DNA into the right order. So they used 'reference genomes'. They took the sequences of two modern genomes that had already been sequenced – the human and the chimp – and used them as comparisons. Armed with this information, a clever computer algorithm was then able to piece the waif-like scraps of ancient DNA back together again. In 2009, Pääbo announced to the world that he had finally sequenced the Neanderthal genome.

The sequence was what geneticists call a 'draft' genome, meaning that although it was the best they could do with the technology they had, it wasn't quite complete. It may seem odd, but in the world of genomes it's completely acceptable to publish draft versions. A working draft of the human genome was published in 2000, three years ahead of the finished article. But I know of no other research areas where the practice is embraced. The team at CERN didn't publish half the Higgs Boson, nor did mankind celebrate when *Apollo 11* got halfway to the moon. But credit where it's due. Pääbo had dedicated his career to overcoming the technical hurdles that had plagued the field of ancient DNA since its inception, and in the process had achieved the seemingly impossible. It was the first genome of an extinct human ever to be decoded.

Make Me a Caveman

In the months following the genome's publication, several things happened. Pääbo's team continued to sequence Neanderthal DNA with a view to upgrading their genome from 'draft' to 'complete', while other researchers got justly excited and started to use the freely available sequence data. Pääbo started to receive letters from men who thought they were Neanderthals, and from women who thought their husbands were. Meanwhile, the *New York Times* reported that a Neanderthal could be brought to life with existing technology for around $30 million.

This claim was attributed to geneticist George Church, and if anyone should know what genetics is capable of, it's him. One of the founders of the human genome project, he helped develop the modern sequencing methods that made reading the Neanderthal genome possible in the first place. Today he continues to push the boundaries of what genetics can achieve. Members of his lab at Harvard Medical School are tinkering with DNA in order to do all sorts of things, including helping to understand human disease, devising new therapies and bringing extinct species back to life. He's currently working on a plan to resurrect the woolly mammoth (see Chapter 3).

'Making a Neanderthal is technically possible,' he tells me when I speak to him on the phone one summer evening, 'if anyone really wanted to do it.' Three years after Pääbo published the draft Neanderthal genome, the complete version was made available. This time, the researchers had dotted all the i's and crossed all the t's*. In fact, every position in the genome had been read an average of 50 times over, making it of similar quality to the human genome.

But it's one thing to reconstruct the Neanderthal genome inside a computer, and quite another to somehow magic it into life. As it exists today, the Neanderthal genome is information stored in digital form. If someone was to print out all of the letters in one single long line, it would stretch for 3,100 miles, the distance from London, England to Boston, Massachusetts. If someone were to publish it in book form, it would fill 5,000 paperbacks and have as much punctuation and plot as chick lit. If you read one letter per second, you'd lose 95 years of your life. A would-be Neanderthal-maker would have to turn this lengthy digital recipe into biological information in the form of actual DNA inside an actual cell.

When the letters of the Neanderthal and the human genome are lined up next to each other inside a computer, it's obvious to see that the vast majority of the letters are the same. Modern humans and Neanderthals, we now realise, share well over 99 per cent of their DNA. So there's no point making a Neanderthal genome from scratch, even if such a thing were possible. Instead, the best option would

---

*OK, so there aren't any i's, but there are t's, c's, g's and a's.

be to start with a human genome inside a human cell and then alter it to become Neanderthal-like.

When I worked in the lab in the early nineties, altering or editing DNA was an arduous labour of love in which only one genetic change could be tackled at a time. But things are different now. Technology has progressed and gene editing is now faster, cheaper and more accurate than ever before. To Neanderthal-ise a human genome, around 10 million changes would need to be made – a tall order but not impossible. It could be done by a process called Multiplex Automated Genome Engineering (MAGE), which can make hundreds of changes to a genome all at the same time, effectively accelerating evolution and earning the technique its alternative moniker of the 'the evolution machine'. It's the brainchild of none other than George Church, the man who once told de-extinction champion Stewart Brand, 'I don't just read DNA, I write DNA.' Church envisages that the process could have myriad benefits, such as making bacteria that churn out biofuels. But he also tells me that although it's not something he plans to do himself, it could also be used to Neanderthal-ise a human cell. The technique would involve breaking up the human genome into thousands of smaller, more manageable chunks, ferrying them into specially adapted bacteria where the desired changes would be made, and then reassembling the modified DNA into 23 pairs of chromosomes inside human cells. The DNA blueprint inside those cells could then be used for cloning. 'Neanderthal' DNA could be injected into a human egg that had had its own nuclear genome removed, then the developing embryo transferred into the womb of a human surrogate who would hopefully be able to carry the embryo to term.

So leaving aside, for one moment, the issue of whether or not this experiment *should* be done, let's think about what the outcome would be. It'll lend weight to the 'should we?' argument that follows. Let's, for the sake of expediency, presume that the baby is male (but it could equally be female) and let's, for the sake of argument, call him Hermann, after the nineteenth-century German anatomist Hermann Schaaffhausen, who studied the original Neander Valley Neanderthal.

Just like all of us, Hermann would be a product of nature *and* of nurture, influenced by his genetic legacy *and* by the environment in which he grew. Certain characteristics, such as height, hair and eye colour, would be strongly determined by genetics, while others, such as personality and preferences, would be more influenced by upbringing. An infant Neanderthal born today would be subject to the same yin and yang of interacting influences that any child is, making predictions at best speculative, at worst hopelessly inaccurate. But bear with me. Let's examine the relevant information, mined both from the fossil record and the Neanderthal genome, to suggest how Hermann might turn out.

Let's start with the birth. Studies of Neanderthal skulls reveal that their brains were around 10 per cent bigger than our own, enough to put even the most experienced of human mothers off a natural birth. As the first Neanderthal born in 40,000 years, midwives and medics would not want to leave anything to chance, so baby Hermann would very likely be delivered by C-section. To all intents and purposes, the freshly delivered baby would look pretty much like any other human newborn – like a grumpy old man deprived of cake. Chances are that Hermann would be pasty faced, maybe freckled. Peoples that have lived in higher latitudes for extended periods have adapted to those conditions by losing the darker skin pigmentation that is common closer to the equator. And when ancient DNA expert Carles Lalueza-Fox from the University of Barcelona, Spain, probed the sequence of a gene related to skin and hair colour, he found that some Neanderthals carried a version likely to result in pale complexions and fiery locks. Hermann might well be a redhead.

But being ginger and/or Neanderthal is no reason to deny any child its basic human rights. They may have been a different species of human to us, but Neanderthals were still human. With this in mind, Hermann would deserve the same upbringing as any human child. We know from fossil finds that Ice Age Neanderthals lived in small, tight-knit social groups where they shared their daily lives and looked after one another – just think of Nandy. Ensconced in the security of a loving, modern family, Hermann would probably thrive. 'A Neanderthal born today would be well suited to the nuclear family,' says anthropologist Thomas Wynn from the University of Colorado, who considers this scenario carefully in his book *How to Think Like a Neanderthal*. 'He would probably be a very loving child.' He would be happiest at home with his immediate family, and would be empathetic and supportive towards his siblings. Outside of the family, however, he might be quite a shy child, mistrustful of those he didn't know. In order to survive, Neanderthals would have had to treat strangers – Neanderthals from neighbouring territories or modern humans migrating from Africa – with suspicion and sometimes hostility. So he might struggle to make friends outside of the family group, and find nursery daunting. 'He would almost certainly not want to be left,' says Wynn, 'to the extent that home schooling might be the best option.'

On weaning, an adoptive parent would be wise to watch the dairy and go gluten free; this baby's genome evolved several hundred thousand years before the agricultural revolution, so Hermann would have difficulty stomaching a modern diet heavy in grains and dairy. Despite their image as mammoth-hunting, bison-stabbing carnivores, there is now plenty of evidence to suggest that Neanderthals ate a rich, varied diet that was both organic and locally sourced. In northern Europe they ate a lot of meat, in southern Europe a lot of seafood, including seals, dolphins and shellfish. Analyses of Neanderthal poop and teeth reveal they also ate vegetables that they probably roasted on open fires. Carles Lalueza-Fox has shown that Neaderthals also shared with us a gene that gave them the ability to taste bitter compounds in certain foods. Hermann would do well on a varied modern human diet light in dairy and wheat, but he'd be just as likely to catapult his broccoli across the kitchen as any other child.

Initially, it's probable that family and friends would notice little difference socially, cognitively or physically between Hermann and any other *Homo sapiens* siblings. The classic Neanderthal features – the sticky-out brow ridges, the odd-shaped head and stocky frame – would lie within the normal range of human variability. Modern humans, after all, come in all shapes and sizes, and just as some of us look more Neanderthal-like than others, so too a de-extincted Neanderthal could look more or less human. Hermann might not have much of an obvious chin, he might well be stockier and stronger than the average infant, but he'd look no more out of place in a crowd than any other child.

Let's Talk Neanderthal

As he turned into a toddler, Hermann would start to learn to talk. A constant source of argument among anthropologists, there are those who think Neanderthals incapable of language and those who believe that language evolved around half a million years ago, in the presumed common ancestor of Neanderthals and modern humans, the tool-using, fire-making *Homo heidelbergensis*. If these very early humans had language, which is what Dan Dediu, from the Max Planck Institute for Psycholinguistics, Nijmegen, Netherlands, and others have argued, then Neanderthals must have had it, too.

Certainly the requisite biology appears to be there. Their face and throat was put together in such a way that Neanderthals were, in theory at least, capable of speech. Just like us, they had a tiny horseshoe-shaped bone at the back of their throat. In modern humans, the so-called hyoid bone supports the tongue and helps control speech. The Neanderthal hyoid not only looked like ours, it seems to have worked like ours, too.

Aspects of their brain anatomy were also similar to our own. Although brains are generally too mushy to survive fossilisation, the skulls that contain them are not. By studying the inside of the skull, researchers can gauge the overall size and proportions of the brain it once housed. One region, known as Broca's area, is of particular interest. Located in the lower left frontal lobe, Broca's area controls the production of speech. Damage to the area, from a tumour or stroke, for example, can rob a person of their ability to talk. Taking the overall size of the brain into account, our Broca's region is bigger than that of our mute non-human primate relatives, and the brains of Neanderthals were, it seems, similarly proportioned – evidence perhaps that Neanderthal brains were capable of generating speech.

Then in 2007, Svante Pääbo and colleagues showed that Neanderthals carried a human-like version of a gene related to speech and language: the *FOXP2* gene. People with an abnormal version of this gene have speech impediments, learning difficulties and struggle with certain kinds of grammar, but Neanderthals have the normal sort – evidence again for the ability to talk?

Collectively, the findings are an enticing smorgasbord of titbits, but they're not a smoking gun for witty repartee. The physical articulation of speech involves far more than the hyoid bone. In non-human primates, and perhaps in Neanderthals, Broca's area is used for non-linguistic tasks like reaching with the hands. And *FOXP2* is not a gene 'for language', it's a gene related to language. It also does other things. On sober reflection, the evidence is circumstantial and, some would say, tenuous. Language, unfortunately, does not preserve directly in rocks nor even in the genome, but if one takes a step back and looks at the cumulative picture from 150 years of Neanderthal research, the picture that emerges is of a people that had to be able to communicate with a level of sophistication way beyond inarticulate grunts and gesticulation.

These were people who made some complicated bits of kit: distinctive curved flint knives intentionally dulled on one side so they could be handheld; foot-long ivory spear points with incisions to help them fit onto wooden handles; and 'lissoirs', polished tools hewn from deer ribs, used to smooth hides and create softer, more water-resistant leather. That they could learn to make these things by observation alone is unlikely. More probable is that they received some verbal instruction. These were people who were expert hunters. They relied on a detailed knowledge of the local landscape and of hunting tactics to locate and bring down their prey. It takes more than one man to fell a mammoth, so it's likely that Neanderthals communicated with one another to discuss hunting strategies and coordinate their kills, and used language to pass their knowledge on through generations. That they made ornaments from teeth, bone and shell, and prepared pigment and painted their bodies with it, implies they were capable of symbolic thought – a prerequisite, some argue, for language where abstract words or symbols acquire reliable meaning.

Multiple lines of evidence, circumstantial though they may be, cumulatively suggest that Neanderthals were capable of speech. 'I may have no direct proof, but I have a strong suspicion that Neanderthals had language,' says Thomas Wynn. 'They're too closely related to us not to have some sort of language.' With the same genome instructing development, and the right environmental cues, a modern-day Neanderthal should be similarly able to acquire and generate language.

As Hermann grew into a toddler, he'd start to babble and then to talk. His first word could be 'mama', it could be 'mammoth', but in the early days at least, there'd be little difference between Hermann's verbal skills and those of any other human infant. It's only a little later when kids start constructing and understanding more complex sentences – 'I know you want to throw the broccoli at your sister but you know that I don't want you to' – that Hermann might start to lag. Such complicated sentences could prove hard to understand and hard to construct. Ironically, given it's the country where the first Neanderthal was found, the German language, with its verb-dense sentence endings, could prove particularly difficult for Hermann to master. But just like his appearance, Hermann's ability to master language would still fall within the normal range of modern human variation. There are, after all, plenty of us who prefer to avoid over-complicated, unnecessary syntax and keep things simple.

Another subtle difference could be in the way that Hermann sounds. Neanderthal vocal tracts were shaped slightly differently to our own, prompting speculation that their voices may have sounded different. In 2008, anthropologist Robert McCarthy from the Florida Atlantic University in Boca Raton simulated Neanderthal speech based on reconstructions from fossils. It was the first time anyone had heard a Neanderthal voice – albeit one generated by a synthesiser – for 40,000 years. He found that the ancient humans were unable to generate certain sounds, known as quantal vowels, which are present in modern speech and help people with different-size vocal tracts to understand each other. Neanderthals, McCarthy concluded, would have struggled to produce the subtleties of the sound 'ee'. So Hermann's adoptive parents might be unsure whether the child wanted to 'heat' or 'hit' his siblings.

Intelligence-wise, Hermann would probably be no Einstein, but neither would he be the halfwit his Neanderthal forebears were erroneously purported to be. On the negative side, Neanderthal technology changed little over hundreds of thousands of years, suggesting that Neanderthals weren't natural innovators. Fifty thousand years ago, Neanderthals in Europe were still killing large animals the way they always had: up close and personal with big, stabby spears, while the new kids on the block, modern humans from Africa, had invented spear throwers so they could kill more safely, from a distance. It's evidence, some say, that Neanderthals were lacking in working memory, the ability to actively hold and manipulate multiple pieces of fleeting information in the brain. And as working memory correlates with intelligence it is, by extension, an indication that Neanderthals were not the cleverest. But on the plus side, Neanderthals did survive for hundreds of thousands of years despite a constant battering from some of the most erratic and extreme weather the world has ever seen, making them resourceful and adaptive. They might not have passed exams or mastered algebra, but that doesn't make them stupid. They may not have been clever in the same way we are now, but to deny them the same potential would be, in my view, a mistake.

So what have we learned about Hermann? The picture emerging is of a shy, socially awkward child that looks and sounds a bit different, but that is loving and happy at home and more or less hitting his developmental milestones within the normal timescale. In short, Hermann would be pretty much like millions of other kids alive today.

His fondness for family life would continue into adulthood. If Nandy is anything to go by, he'd probably stand by his parents and look after them in their old age. He'd probably want children of his own and one day he'd meet a girl and settle down. He'd be a pragmatic individual who'd put his family first and work hard to provide for them. A man of few words with a tendency for straight talking, he'd excel in a job that played to his pragmatic, no-nonsense nature. Thomas Wynn suggests several careers that would fit. A natural craftsman, capable of reiterating complex learned procedures, he'd make an excellent blacksmith or machinist. He'd also be good at fixing up cars or repairing engines, but he wouldn't be restricted to engineering. Neanderthals had a deep understanding of the natural world, carried detailed mental maps of their environment, and planned and executed audacious hunting trips. So Hermann might make a skilled commercial fisherman or excel in the military. He might not be able to explain the intricacies of GPS or military radar, but he'd certainly be able to use them. He might not come up with innovative problem-solving strategies, but he could certainly implement learned manoeuvres, and he'd be a determined and fanatically loyal co-worker.

Ice Age Neanderthals endured extreme hardship, were frequently injured and learned to tolerate pain, so Hermann would be unlikely to succumb to that most dreadful of modern scourges, 'man flu'. He'd also, given the story of Nandy, be kindly and sympathetic towards his partner if she ever became ill, say, the morning after a night out with her friends. He'd still struggle with strangers, but once he determined the aliens were friendly, he would be warm and generous, perhaps overly so. Hermann could be naïve and gullible and easily conned. He'd probably be no good with money.

The picture we have is of a man that talks little, moans less and nurses a girl with a hangover. He loves his family and steadfastly brings home the bacon but relies on his wife to manage the purse strings. If you're a little like me, Hermann's starting to look like a bit of a catch. The only possible drawback to marrying a Neanderthal, speculates Thomas Wynn, is that he might want to be with you physically 24:7. 'It seems "his and her spaces" were not a Neanderthal forté,' says Wynn. If you like needy men, however, he could be the one for you.

In the Ice Age, their arduous lifestyles left adult Neanderthals struggling with arthritis and they rarely lived beyond 40. But there's no reason to think, with the added comforts of modern life, that this Neanderthal wouldn't live arthritis-free, well into his eighties or longer. Hermann and his wife could grow old together.

Perhaps We Could, But Should We?

So there it is – the story of a Neanderthal born into a twenty-first-century world. It starts with ancient DNA, harvested from the remains of genuine Neanderthals long dead. It then sees scientists use this Ice Age recipe to 'Neanderthal-ise' a modern human genome, and then use cloning to create a living, breathing Neanderthal baby that would live and die in our busy, chaotic modern world. It would be an incredible feat.

As it stands today, there are no laws to prohibit the cloning of a Neanderthal. 'It's not illegal to do what you're talking about,' says bioethicist Arthur Caplan from the NYU Langone Medical Center in New York. Around 10 years ago, Caplan sat on the advisory committee to the United Nations on Human Cloning, a group that sought to reach an internationally binding agreement on reproductive and therapeutic cloning. In the end, countries reached their own legislative solutions, but back then ancient human DNA was so far off the radar as to make cloning a Neanderthal unheard of. 'No one was thinking about anything other than *Homo sapiens*,' says Caplan.

But just because we *could* do it doesn't mean we should. Cloning is a risky business. We know that cloned animals often die in the womb, and that if they are born at all they sometimes have physical deformities and live unnaturally short lives. There would probably be many deformed and dead Neanderthal babies long before a healthy one comes along. Their resurrection through cloning puts the lives and wellbeing of surrogate and baby at stake … to what end? No woman should ever have to endure the uncertainties, the psychological and physical trauma of carrying another human species to term. The risks are just too great. It's with very good reason that human reproductive cloning is illegal in over 50 countries.

In 2006, Francis Collins, director of the Human Genome Project, argued: 'Scientists, ethicists, theologians and lawmakers are essentially unanimous that reproductive cloning of a human being should not be undertaken under any circumstances … Implanting the product of human SCNT [cloning] into a uterus is profoundly immoral and ought to be opposed on the strongest ethical ground.' Others have called it 'criminally irresponsible'. If we recognise, as we must, that Neanderthals were basically human, related closely to us in evolutionary time, then we must award them the same moral status, rights and privileges that we grant ourselves. Neanderthals should not be cloned.

Researchers have pored over the remains of more than a hundred Neanderthals, scattered across time and geography. From them, we know more about Neanderthals than any other extinct hominin. We know where they lived, when they lived and how they lived. We know what they ate and how they hunted. From the fossil and genetic record, we can even make educated guesses as to their behaviour, personality and thought processes. We know what they looked like, right down to their skin and hair colour. We can use computer modelling to put flesh on their bones – we don't need to do it for real. We can study their genes by adding them into cultured cells, we don't need to bring back an entire individual. There is little to be learned from a de-extincted Neanderthal person that we don't already know.

In the meantime, anthropologists continue to argue passionately over the interpretation of Palaeolithic fossils, and the debate over how sophisticated Neanderthals were continues to swing back and forth. While it's obvious that Neanderthals weren't as cultured or erudite as our modern selves, we're in no danger of reaching a consensus on how they compared to Ice Age modern humans any time soon. Nor can we agree on why they went extinct. But Hermann couldn't help with any of that. He wouldn't have the answers. This Neanderthal would be very different to those who battled the elements tens of thousands of years ago. A product of nature and of nurture, his DNA might be similar, but Hermann would be growing up in a very different world to the one his ancestors evolved in. He'd be exposed to fast cars and fast food, social media and smart phones. He'd grow up in a time of consumerism, global superpowers and boy bands. He could never be the mammoth-hunting caveman of the past. Anyone hoping for some kind of loin cloth-wearing freak show would be disappointed to find a twenty-first century Neanderthal reassuringly normal. If you were sitting next to him on a park bench, you'd probably never raise an eyebrow. Hermann, as I hope I've convinced you, would be so average as to blend in with the other seven billion humans on the planet.

A subject of intense research efforts, we've also learned recently how the environment can influence gene activity via a process known as epigenetics. What we eat, drink and smoke, our social status and even the care we receive as children has the potential to permanently alter the way our DNA works. Hermann might have an Ice Age genome, but it would be influenced by his twenty-first-century upbringing. His patterns of gene activity would be different to those of his genuine Neanderthal ancestors. As a result, he might look or behave differently, or perhaps succumb to different diseases. Genetic puritans (or 'molecular biologists', to give them their proper name) will tell you this Neanderthal would not be the real deal. His would be a genome out of time. The only way to find out what Neanderthals were really like would be to build a TARDIS and travel back forty thousand years or so.

Have I convinced you yet that there is little to be learned from bringing a Neanderthal back to life? It's my strong belief that just because we could de-extinct a Neanderthal doesn't mean we should de-extinct a Neanderthal. It's precisely because they are so similar to us that this experiment should never be allowed to happen. Resurrecting a Neanderthal, or any other extinct human species, is a line that should never be crossed.

No Sex Please, We're Neanderthals

But maybe we don't need to bring the Neanderthals back. Maybe their legacy is still with us. When Svante Pääbo decoded the Neanderthal genome, he compared it with the genomes of modern humans living in different parts of the world today. What he found surprised everyone. To that point, all the evidence had suggested that Neanderthals and the modern humans they lived alongside had kept their relationships platonic. But Pääbo's results suggest something much more intriguing. The genomes of people from Europe and Asia contain traces of Neanderthal DNA. Tens of thousands of years ago, Neanderthals and modern humans interbred, creating children and a genetic legacy that still exists today.

Others, including geneticist Joshua Akey from the University of Washington, have since confirmed the findings. People of European and Asian origin carry between one and four per cent Neanderthal DNA, while people from Africa carry none. The implication is that, after they left Africa, modern humans bred with Neanderthals. The humans that remained in Africa never had the chance to meet a Neanderthal, much less invite them back to their cave, so their genomes are Neanderthal-free.

Although the total amount of Neanderthal sequence in any modern human is relatively low, the cumulative amount of Neanderthal DNA that persists across all humans is a staggering 20 per cent. From his data, Akey estimates it took around 300 successful 'hybridisation events' to end up with the pattern of Neanderthal DNA seen in living people today. Of course, modern humans and Neanderthal may have had sex more often than this. 'Maybe at the time, modern humans didn't find Neanderthals as different as we do today,' says Akey, but their trysts may not always have resulted in viable pregnancies. 'They may have been on the fringes of biological compatibility,' says Akey.

That vast swathes of the cultured, educated, shaven, Western world are part-Neanderthal may come as something of a shock. You may not know it, but you may also be part-Neanderthal. But you don't have to take my word for it.

Anyone can take a test to find out how much of their genome is of Neanderthal ancestry.

And so it came to pass that one sunny, summer evening I announced to my husband that I was going to do a DNA test to find out how Neanderthal I am. 'Do you really need to take a test to work that one out?' he replied. He's a funny man. Several days later, as the shoe-shaped bruise on his *cojones* began to fade, I broached the subject again. 'Fine, fine, do whatever,' he said quickly. For $99 (plus postage and packing), you can send your spit to a US company that will analyse your DNA and help unravel your ancestry, including your genome-wide percentage of Neanderthal DNA. Outside, our three children were taking it in turns to pin each other down, swing from trees and catapult chickens off the trampoline. What if I really did carry Neanderthal genes and had passed them on to my children? Could it explain their rambunctious behaviour, their instinct to brawl and blatant disregard for animal welfare?

I registered online and paid my money, then four days later, the postman delivered a parcel; a box in a padded envelope in a padded envelope … because you can't be too careful when it comes to protecting your DNA. The box was the size and shape of a disappointingly proportioned chocolate selection, but it had the words 'Welcome to you' printed on it and lots of colourful chromosomes emblazoned on the lid. Inside was a high-tech tube and an instruction leaflet that advised me to stop eating and drinking for 20 minutes then spit into the tube. It was the longest 20 minutes of my life. Walking away from the biscuits and endless cups of tea that supplement my working day was arguably one of the hardest things I've ever had to do. Fortunately, spitting into the test tube was much easier, as was popping the tube shut, shaking it up a bit and slipping it back into the pre-paid envelope that carried it all the way back to California and the shiny labs of DNA testing company '23andMe'.

Two weeks later an email dropped into my inbox to tell me that my initial report was ready. Like a teenager waiting for an exam result, I nervously logged on to the company website. The results were in. First up, a few things that came as no surprise but that reassured me the results were genuine. My DNA is 99.9 per cent European, of which I'm 53 per cent Northern European and 38 per cent Eastern European. My dad was from England, while my mum is from Lithuania, so that sounds about right. Then I have a dash of Ashkenazi (1.4 per cent) and Southern European (3.8 per cent), and an even smaller smattering of East Asian and Native American (0.1 per cent). Perhaps there are some family secrets hiding in the boughs of the Pilcher family tree. Next up came the news that I have one new second cousin, 171 fourth cousins and 817 'distant cousins' that I never knew I had. Suddenly the Christmas card list is looking daunting.

And then came the moment of reckoning. The average 23andMe punter, the website tells me, has a genome containing 2.7 per cent Neanderthal DNA. But my genome contains 3 per cent Neanderthal DNA. I tell my husband. He retreats to a safe distance then says, 'Is that all?' This means I have 0.3 per cent more Neanderthal DNA than the average person, which puts me in the 91st percentile for Neanderthal-ness. I am officially a cavewoman.

I'm not sure what to make of it. The truth is that no one knows exactly what all of the Neanderthal DNA that persists in today's modern humans actually does. There are some areas in our genomes where there is little or no Neanderthal DNA, and others where there is more than expected, evidence that some sequences were advantageous and so positively selected for, while others conferred no benefit and were lost over time. Single genes influence many different things, so the DNA I've inherited will probably have diverse effects. Many of the Neanderthal genes that live on in people today, studies suggest, appear to influence skin pigmentation and keratin, a protein found in skin, nails and hair … which could explain why I have the hairstyle of Stig of the Dump, which I've been insisting for years is genetic.

Suddenly it all begins to make sense. I can never find a hat that fits. I have hairy toes. There isn't enough wax in the world to tackle my bikini line. My children, also part Neanderthal, don't just get their spirited behaviour from me and my husband. We can pass the buck to some prehistoric person who spent his days chasing mammoths and his nights sleeping in caves. It can explain why they like camping outdoors, playing with fire and hitting things with sticks. This is excellent, excellent news, especially if they turn out to be as benevolent as their Nandy-caring antecedents.

I'm pleased to be more 'caveman' than most. For too long, Neanderthals have been hapless stooges. They've been maligned and misunderstood. But perhaps now, with the discovery that so many of us are a little bit Palaeolithic, things might start to change. We don't need to resurrect them – we should never resurrect them – to know that Neanderthals were resourceful and resilient, compassionate and caring. They lived sustainably in their Ice Age world for hundreds of thousands of years without wrecking it, while in the comparably short time we've been around, us modern humans have made quite a mess. Perhaps it's time the word 'Neanderthal' stopped being used in the pejorative and started to be used as a compliment. I say it loud, I say it proud: I'm honoured to be part Neanderthal … and I will never wax my legs again.

CHAPTER THREE

King of the Ice Age

Malyi Lyakhovski is a tiny, remote island lodged in chilly Siberian waters. Snow-covered and ice-locked most of the year, no one lives there. Any visitors that do set foot on its sparse, treeless tundra must come well prepared; wrapped up warmly and with rifles. In winter, temperatures drop to 30 below, and there are hungry polar bears. But come they do. Every summer, when the feeble sun clings to the horizon, a unique band of hunters make their way from the Siberian mainland to the little island, but they're not looking to make a kill. They come to scavenge the remains of animals long dead; the bodies of giant beasts that perished tens of thousands of years ago. For entombed in the permafrost are the frozen remains of the ultimate Ice Age giant, the woolly mammoth (*Mammuthus primigenius*).

When explorers first stumbled across Malyi Lyakhovski in the eighteenth century, they thought the island was literally made of bones and tusks. What can be said with certainty is that the archipelago and the Siberian mainland it belongs to are one massive mammoth graveyard, littered with the remains of these fallen behemoths. As our climate changes and the planet warms, the frozen north is thawing faster than ever, and mammoths, or parts of them, are being discovered with increasing frequency. Their appearance is luring keen-eyed chancers, some of whom are prepared to risk their lives in pursuit of their quarry. Mammoth-hunters have been known to scuba-dive in freezing Arctic waters, prod around inside unstable ice caves and use pressurised water jets to blast bones from crumbling cliffs. Such are the rewards that the ends, apparently, justify the means. A huge, spiralling mammoth tusk can earn its finder tens of thousands of dollars, and there are millions of bones to be found. It's a modern-day gold rush, not in precious metal but in body parts. In a region turned ghost town after the fall of communism and the closure of Soviet-era mines and factories, mammoths provide an economic lifeline to intrepid, cash-strapped locals, and a new word has entered parlance: *'mammontit'* – 'to mammoth', or to go bone hunting. But as the easy-to-reach mainland areas are picked clean, foragers are having to venture to ever more remote locations – like Malyi Lyakhovski.

It was on one such expedition in August 2012 that tusk collectors chanced upon the find of a lifetime. Standing on a wind-battered hill not far from the island's north-east coast, they happened to look down and glimpse contours of bone peeping through the tundra. Framed against a backdrop of scraggy grasses and moss were a few fragments of skull and a shard of tusk. They stooped to take a closer look, feeling the finds gently with their gloved hands … then noticed something unmistakeable. Lying across the tusk, camouflaged against the dirty ground, was a tousled, brown trunk. Unmistakably mammoth, the curved, tentacle-like structure was covered in a matt of fur and had at its tip the tactile 'fingers' once used by the animal to help it feed. To find bones and tusks is not uncommon, but to find flesh and tissue is very rare. Recently exposed, the melting, semi-frozen structure had yet to decay or attract the attention of scavengers. For the hunters, it was pure gold. If the trunk had somehow managed to survive, perhaps, they reasoned, other body parts could also be found in the frozen ground. So they set to, hacking at the earth with their picks and their spades, only to find they were standing on the grave of what seemed to be an almost complete animal. What lay on the surface was just the tip of one enormous frozen mammoth Popsicle.

The weather, however, was against them. The winter winds were closing in fast. Exhuming a mammoth would take time and there was no way the hunters could excavate the body before blizzards forced them to flee the island. So they returned to the mainland empty-handed, where they pondered their next move. They could keep quiet, retrieve the corpse the following spring, then make a fortune by selling it privately. Or they could alert the local scientists at the North-Eastern Federal University (NEFU) in Yakutsk, Siberia, who were offering a more modest reward to anyone that could lead them to the remains of a well-preserved mammoth. It meant the difference between the Malyi Lyakhovski mammoth ending up in private hands or being made public through the research of the scientists and the publication of their finds. Fortunately for us, the tusk-hunters chose the latter. They alerted a man called Semyon Grigoriev, head of the university's Mammoth Museum, who was more than a little excited. In his time, Grigoriev has handled the remains of many mammoths, but with its fleshy trunk, this one promised something really special – an unprecedented level of insight into the mammoth's biology and life story and, more controversially, the chance to bring the animal back from the dead.

Frustratingly, Grigoriev and his team had to wait an excruciating nine months for the Arctic winter to subside before they could travel to Malyi Lyakhovski. In April 2013, they made the bumpy schlep, hundreds of miles across the snowy wastelands of northern Siberia and the frozen Laptev Sea to the remote island where the mammoth still lay. GPS coordinates lured them to the exact spot where the tusk-hunters had stood less than a year ago. They too began to chip away at the frozen ground, excavating a trench, 1.8m (6ft) deep around the frozen carcass. With the body revealed in its entirety they were able to stand back and realise the scale of their find. The mammoth was … mammoth. Looking down on it from above, they could see that the top half of the body had been gnawed at, so that it looked like a huge lump of gnarled meat. But from inside the trench, the rest of the body looked remarkably intact. Three legs, its torso, most of its head and trunk were still there. But instead of being desiccated and dry like most other mammoth bodies found before, in parts this one had flesh on its bones, skin on its flesh and fur on its skin.

But there was one more surprise in store. When the scientists prodded the creature's belly with a pick, a dark brown liquid came oozing out. It looked like blood, but how could a mammoth, frozen in the ground for tens of thousands of years, still have blood that flowed?

A Woolly Legend

Woolly mammoths are the undisputed Kings of the Ice Age, the geological era also known as the Pleistocene. They evolved in the midst of this interminable cold spell several hundred thousand years ago, at a time when immense ice sheets covered swathes of the northern hemisphere, locking in so much water that they created cloudless blue skies. Underneath that sky lay open, fertile grassland that covered much of northern Eurasia and North America: the mammoth steppe, home to *Mammuthus primigenius*. Perfectly adapted to life in the freezer, with their enviably thick, lush locks, vast herds* of mammoth grazed away to their hearts' content.

Then, little by little, towards the end of the Pleistocene their numbers started to diminish. No one really understands why. Some blame human hunting, some climate change, others a bit of both. Whatever the reason, they disappeared from Siberia around 10,000 years ago, and from their final hiding place, a northerly island called Wrangel, as recently as 3,700 years ago – it is incredible to think that woolly mammoths were still alive at the time that Ancient Egyptians were setting up home in the Nile valley and dreaming of pyramids.

Gone, but not forgotten. Immortalised in cave paintings by the early humans they lived alongside, reconstructed by us from the bones and body parts they left behind, we understand more about these shaggy beasts than any other extinct prehistoric animal. If we were to choose just one animal to bring back, it had better be one we know a lot about.

Mammoths are everything that you'd want in a de-extincted animal. Vegetarian – so they won't eat you. Iconic. Inspiring. Larger than life. Their enormous, curvy tusks could grow to over four metres (13ft) long. A thighbone could be more than a metre (3.3ft) long, and a single tooth the size of a sliced loaf. Their sheer scale almost defies imagination, so much so that we've been concocting stories to explain them for millennia. In the past, some thought their bones belonged to a lost race of giants, others that they were a rare breed of underground rodent, which died when exposed to the sun. Western scholars once thought the biblical flood of Genesis had swept mammoth bones to Siberia, while others believed the remains

---

*I'm presuming that 'herd' is the correct collective noun for mammoths. If early modern humans or Neanderthals ever possessed such a word it has been lost in the mists of time. Perhaps a 'ball' of woolly mammoths would be better.

belonged to elephants that had strayed from the herds of Alexander the Great. With a central hole in its skull where the trunk would have attached, some think mammoth fossils inspired the legend of the one-eyed Cyclops. Even today, people are wary. Some locals refuse to disturb newly found mammoth remains for fear their actions will bring bad luck, like a mummy's curse. We may think ourselves a little more mammoth-savvy these days, but there's something about this animal that gets our imaginations running on overtime. Mammoths may be long gone, but we still incorporate them into our culture – think Manny from *Ice Age*, Mr Snuffleupagus from *Sesame Street* – and such is their stature in popular culture that their very name has become synonymous with enormity, vastness and greatness. We've all heard of them, we've all seen pictures of them. Who wouldn't be intrigued to see one in the flesh? Bringing back a butterfly, a frog or some other smaller creature might be nice, but it would be easily overlooked, literally. A resurrected mammoth, on the other hand, would be a showstopper.

There may even be several fairly staggering economic and ecological reasons to rush forward their revival. In their day, woolly mammoths were the Alan Titchmarsh* of the Arctic. They trundled around eating grass, trampling saplings and fertilising the ground via their nutrient-rich dung. But when they disappeared, things changed. Jacquelyn Gill from the University of Maine has analysed pollen, charcoal and spores from North American sediment

---

*For those who don't know, Alan Titchmarsh is a legendary British gardener and national treasure. I'm not suggesting that he eats grass, tramples saplings or shits in his own back garden. But he does sculpt landscapes with his green fingers. It's also hotly tipped that when Prince Philip pops his clogs, the Queen may well marry Titchmarsh and let him put raised beds in the grounds of Buckingham Palace.

cores to reveal that, over the next few thousand years, the landscape altered dramatically. Without mammoths to knock them down, temperate deciduous trees such as elm and ash sprung up next to cold-loving conifers like larch and spruce. And without mammoths to 'mow the grass', plant litter began to build up. Every couple of centuries, fire ripped through the forests; in time the lush grasslands of the 'mammoth steppe' were transformed into unproductive mossy tundra. Cue tumbleweed blowing across the page.

Today, the remains of that rich, fertile Ice Age ecosystem are locked up inside the Arctic's frozen soil. The permafrost is thought to harbour an estimated 500 gigatonnes of sequestered organic carbon; that's two to three times as much as occurs in all the existing rainforests combined. It's a carbon time bomb. As our world warms, the frozen north is thawing and little by little that carbon is being released into the atmosphere as gas. It's warming the world and computer models suggest that ice-free Arctic summers could be with us as soon as 2052, if not before.

But mammoths, some think, could help keep the Arctic cold. 'When big animals graze, they trample the snow, which exposes the surface of the soil to cold air', says biologist Sergey Zimov from Siberia's Northeast Science Station. 'This helps keep the ground frozen.' Zimov has shown that in Siberia, soil temperatures in areas where big animals graze are, on average, several degrees colder than where grazers are absent. Recreate the mammoth steppe, with its characteristic flora and fauna, and the grazers could help keep the carbon time-bomb from blowing.

Zimov, a rollie-smoking, pony-tailed Grizzly Adams of a man with a ZZ Top-style beard, already has the project in hand, and if it all sounds a little *Jurassic Park*, then please allow me to elaborate ... For the last 20 years, Zimov has been trying to recreate the mammoth's original ecosystem in the shape of a nature reserve he calls Pleistocene Park. It's situated in one of the coldest places on earth – in the Sakha Republic, northeastern Siberia, eight time-zones east of Moscow and a 4.5-hour flight away from the nearest city, Yakutsk. Zimov is filling his park not with dinosaurs, but with big imported herbivores including Yakutian horses, reindeer and moose, and with local carnivores, including wolverines and bears. These are the kinds of animals that lived there when mammoths were around. Within one season of being helicoptered in, Zimov found that his stocky Yakutian horses had turned one area of meagre, moss-filled wilderness into a lush, steppe-like grassland – promising results hinting that, with the right mix of animals, the mammoth steppe could indeed be recreated. But there is, of course, one missing ingredient. Mammoths! 'My responsibility,' Zimov told me, 'is to prepare the ecosystem. I will prepare the environment for the mammoths.' If and when the mammoths are brought back, they will have a home already waiting for them in the shape of Pleistocene Park.

Young Dreams

The first scientist to try to de-extinct the woolly mammoth was cell biologist Viktor Mikhelson from Leningrad's Institute of Cytology, back in the 1980s. He was inspired by the discovery of a mammoth calf called Dima, found lying on its side with its trunk curled up near a tributary of the Kolyma River in northeastern Siberia. Desiccated and wrinkled, Dima was covered in straw-coloured 'baby' fur with tufts of darker adult hair just starting to peek through. The calf was so lifelike it almost cried out for resurrection, prompting Mikhelson to wonder if its cells could be salvaged for cloning.

It was a bold idea. At this point in time, no one had ever cloned a mammal before, much less a dead one, much less one from the last Ice Age. Mikhelson planned to take the DNA-containing nucleus from one of Dima's cells and inject it into an elephant egg that had had its own nucleus removed. If the reconfigured egg started to develop he would then transfer the resulting embryo into the womb of a surrogate elephant mum who, if all went well, would then give birth to a cloned mammoth – Dima's living, identical twin.

Mikhelson tried for months but never made it to first base. The thawing corpse had been left in the open for several days after it was unearthed, so had started to rot. Then, after the body was transported to Leningrad, taxidermists botched what was left of it. The carcass that nature had preserved so perfectly for thousands of years was now black, bald and full of chemicals. Mikhelson never stood a chance. If scientists were to resurrect the mammoth they'd have to find a better specimen to clone from.

It would be more than 10 years before anyone tried to resurrect the woolly mammoth again. This time it was the turn of the Japanese. At Kagoshima University, reproductive biologist Kazufumi Goto was working with wagyu, an expensive type of cattle highly prized for its marbled meat. Goto had been collecting semen from bulls then using the samples to make test-tube cattle. The idea was that, through *in vitro* fertilisation (IVF), a single shot from a high profile sire could be used to produce thousands of equally desirable offspring. But Goto realised the same method also worked with sperm that were dead. So, he mused, if a dead sperm from a bull could be used to create new life, why not dead sperm from a frozen mammoth?

His plan was to go to Siberia, find a mammoth, then use its sperm to fertilise an elephant egg. Unlike Mikhelson's clone, which would be virtually pure mammoth, this animal would be half mammoth, half elephant – a hybrid, or 'mammophant', if you like.* If the animal survived and was able to reproduce normally, Goto planned to breed it back with another mammophant to bump up the quota of mammoth DNA in the offspring. The process could then be repeated until an almost genetically pure mammoth had been created a couple of generations later.

*Or indeed, 'elemoth'.

The odds, however, were stacked against him. First, Goto would have to find his mammoth. For obvious reasons, the animal would need to be male, which by the law of averages ruled out 50 per cent of the frozen finds. It would also need to be an adult. Like elephants, mammoths would have become sexually mature at around 10 to 15 years of age, so finding juvenile mammoths would do little to help the cause.

Second, the adult male would have to be exceptionally well preserved. For the dead or dying beast, that would have meant being frozen quickly then covered up. Freezing something quickly means that tissue-munching bacteria have less time to get to work and the carcass is less likely to rot. Covering it up means scavengers can't eat the body and DNA-damaging cosmic rays are less likely to ravage the cell nuclei. To achieve this, the mammoth would have to have met a fairly gruesome death, such as getting stuck in a bog then covered in snow (like Dima), or falling through the ice and drowning in a frozen lake.

But even if that happened, there was no guarantee that viable sperm could be recovered. Where humans, dogs and many other animals dangle their family jewels gratuitously outside their bodies, mammoths, like modern-day elephants, wore them discreetly, on the inside. Located deep inside their bodies their melon-sized testicles would have pumped out sperm that then matured nearby in a coiled duct system. It would have been about the last part of their dead body to freeze. Indeed, it's been estimated that even on the coldest of nights, a six-tonne adult mammoth would still have taken several hours to freeze solid, giving bacteria ample time to get to work on its gonads. The chances of mammoth sperm being frozen quickly enough to avoid damage were as small as the sperm themselves. And that, for the record, is very, very small indeed. Assuming their reproductive biology to be similar to that of modern-day elephants, Goto and his team were looking for cells just one-tenth of a millimetre long. Looking for mammoth sperm would be like looking for the proverbial needle in a nutsack.

Even if Goto struck lucky and found viable mammoth sperm, there was no knowing for sure if the recovered cells would be able to fertilise the eggs of another species, even a closely related one. For egg and sperm to marry successfully, the strands of genetic material or chromosomes from one must align perfectly with those from the other – a feat that is possible if the two species contain an equal number of chromosomes, and problematic if they don't. So lions and tigresses, with 38 chromosomes each, can yield fertile 'liger' offspring. But horses and donkeys, with 64 and 62 chromosomes respectively, tend to produce infertile mules and hinnies. Even when two closely related but different species have the same number of chromosomes, things can be tricky. African and Asian elephants, for example, have the same number of chromosomes, but there has only ever been one hybrid between the two, a male calf called Motty, born the year after Elvis died, in 1978.

For his first trip, in the summer of 1997, Goto invited along a colleague, reproductive biologist Akira Iritani, from the Department of Genetic Engineering at Kinki University, near Osaka, Japan. Iritani came with an impressive CV. In the late eighties he achieved a world first when he injected a single rabbit sperm into a single rabbit egg and made a baby rabbit. Not that rabbits need much help in this department – they breed like rabbits after all – but this was the first mammal to be produced using intracytoplasmic sperm injection, or ICSI, as it is known. Now an established technique in fertility clinics around the world, it has been used to help tens of thousands of couples have children. Iritani also wanted to make a mammoth, but he wasn't interested in sperm. He had a different plan.

Hello Dolly

The summer before they were due to leave, something happened to boost Iritani's spirits. On 5 July 1996, a little lamb was born on a farm in Scotland. To be sure, there were lots of lambs born that day. But this one was special. This wasn't any old lamb. This was Dolly, who went on to become the most famous, most photographed sheep in the world. Mammalian cloning, which had been nothing more than a pipe dream for Viktor Mikhelson back in the eighties, had finally arrived. It was one small step for lamb, one giant leap for lambkind. She was named after busty ballad belter Dolly Parton because the cell used to clone her came from the mammary gland of a female sheep. Her birth was trumpeted in the world's leading scientific journal, *Nature*; she stole headlines around the world, and even bagged the front cover of *Time* magazine. A superstar sheep who played her entire life out in front of the watchful lens of the world's media, she could have been a demanding diva, yet Dolly was reassuringly normal to those who knew her. 'She was a sweetie,' says developmental biologist Michael McGrew from Scotland's Roslin Institute, where Dolly was born. 'She spent a lot of time with people so she became really tame.' But Dolly was more than just a nice lamb to hang out with. Her birth heralded the arrival of a new era. Just as the Gregorian calendar can be split into BC and AD, so too the field of cloning can be divided into BC: 'Before Cloning', and AD: 'After Dolly'. Before Dolly, mammals had never been cloned using DNA from an adult cell. After Dolly, that all changed.

She was important because she demonstrated, for the first time, that genes in the nucleus of a mature, adult cell can be reprogrammed into a much younger, embryonic-like state. 'Old' DNA could be tricked into becoming young again.

'It was a dramatic discovery,' says Iritani. If a sheep could be cloned, he thought, why not a mammoth? It was after all, a mammal too, albeit a very large, shaggy, dead one with a trunk. Iritani planned to pick up where Mikhelson had left off and try to de-extinct the mammoth via cloning. If only he could find an intact, viable cell then he could employ the same basic procedure used to make Dolly to fashion a baby mammoth.

Over the next couple of years Iritani and Goto travelled to Siberia several times to look for mammoth tissue in the melting tundra, but the best they could find was a scruffy bit of rump that looked and smelled like a student's bath towel. And if it wasn't in great condition when they found it, it was even worse after Russian customs stalled its departure. Four months later when the rancid bit of bum finally arrived at Goto's laboratory it was, perhaps predictably, a non-starter. There was no sperm for Goto's hybridisation experiments, nor was there any DNA for cloning. To add insult to injury, they later realised that the sorry sample had probably belonged to a woolly rhino rather than a woolly mammoth. It was all a mammoth fiasco.

A Load of Old Bull

The story then goes quiet for a decade or so. Iritani hadn't given up resurrecting the mammoth. He was just biding his time. Although Dolly had been born, in the late nineties the science of cloning was still in its infancy and the field needed time to mature. Sceptics thought (and still think) that cloning from permafrost-frozen cells was impossible. When cells are frozen for storage in the laboratory, scientists add cryoprotective 'anti-freeze' chemicals that prevent the cells from shattering. In this way, cryoprotected, laboratory-frozen cells can be kept in liquid nitrogen for decades, then slowly thawed and brought back to life. Mammoths, however, simply froze where they fell. With no artificial cryoprotectants to help keep their cells viable, attempts to thaw and grow mammoth cells were fully expected to fail.

But in 2011, Iritani made a bold statement. 'Technical hurdles have been overcome,' he told a British newspaper. 'I think we have a reasonable chance of success and a healthy mammoth could be born in four or five years.'

Iritani's optimism was founded on a couple of methodological advances. In 2008, Teruhiko Wakayama, then at the Riken Centre for Developmental Biology in Japan, and colleagues took cells from a mouse that had been slung in a freezer and frozen whole for 16 years, and used them to clone an entire new animal. The study is all the more remarkable because the cells that were used for cloning were far from pristine. The mouse had been frozen without any form of cryoprotection. So when the cells were thawed, none of them were even intact. The study showed that just because a frozen cell looks in bad shape, it doesn't mean you can't clone from it. Iritani was encouraged.

Then, a year later, a group of Japanese scientists, including Iritani, went a step further when they made clones from a prize bull called Yasufuku. During his life, Yasufuku's sperm were collected and used to sire over 40,000 calves, but when he died (from dementia, not exhaustion) his profoundly productive testicles were removed, wrapped in tin foil and slung in a freezer at -80°C (-112°F). Whether or not it's what he would have wanted we'll never know for sure, but there his testicles lay for over a decade, without cryopreservatives, until Iritani and co. used cells from them to create three identical Yasufuku clones. Even in death, Yasufuku's testicles could do the job. If you could do it for bulls, why not other large frozen animals? 'Our results,' the researchers said, 'suggest the possibility of restoring extinct species, such as woolly mammoths, if live cells can be retrieved from an organ or animal that has been frozen in a freezer or in the Siberian permafrost.' The hunt for the world's best-preserved mammoth was on.

Iritani planned to use a modified version of Wakayama's method to clone a mammoth, and had some promising preliminary data under his belt. In 2009, he published a paper in *Proceedings of the Japan Academy*, where he described the first step in the process. Iritani and his team took cells from a frozen mammoth and injected their 15,000-year-old nuclei into empty eggs. In an ideal world, Iritani would have used elephant eggs. Inside the egg, the mammoth DNA would have started to move around and organise itself into separate, stringy chromosomes. The reconstituted cell would then have started to divide. But this was not to be. Elephant eggs are hard to come by, so Iritani was forced to use mouse eggs as a proxy. It might sound far-fetched, given the obvious size difference of the two animals, but a mouse egg can easily accommodate the nucleus of a mammoth. The hope was that naturally occurring molecules inside the mouse egg might spur the mammoth DNA into action. All in all, Iritani injected more than a hundred mammoth nuclei into more than a hundred mouse eggs, but not a squeak. Perhaps, Iritani surmised, the mammoth DNA was simply too old to behave properly, or perhaps mouse eggs simply aren't capable of reprogramming mammoth DNA. But perhaps, if tissue culture techniques could be improved and better-preserved mammoth cells found, things could be different. 'In the elephant oocyte (egg), the mammoth nucleus might be activated,' he told me.

Iritani is now working on specimens collected from another mummified mammoth. Yuka is an exceptionally well-preserved and complete mammoth; a 39,000-year-old strawberry blonde 'toddler' found lying on its back, legs in the air, in the Siberian permafrost in 2010. The body is of particular interest because it has two large cuts on its back, through which many of its bones have been removed, including its skull, spine and pelvis. There's no way animals could have done this, so the body provides evidence of potential tampering by early humans. What's left, however, is in pretty good shape. In 2012, Iritani travelled to Yuka's home, the Sakha Republic Academy of Science in Yakutsk, Russia, where he signed an agreement giving him access to their mammoths, and collected samples from Yuka. It took a year of Russian bureaucracy before the samples – skin, muscle and bone with marrow – were released, but they're now safely ensconced in Iritani's lab at Kinki University. He's now studying the precious samples with care, on the hunt for that one elusive, viable, intact nucleus that just might help him fulfil his dream. But Iritani and his team aren't the only ones trying to make a mammoth. Around

500 miles away across the Sea of Japan, a controversial scientist is at the forefront of a counter-effort.

From Mutts to Mammoths

South Korean cell biologist Woo Suk Hwang is perhaps best known for a scientific scandal he has spent the last decade trying to forget. In October 2009, Hwang, who claimed to be the first to clone human embryos and create stem cell lines from them, was found guilty of fraud, embezzlement and ethical violations. For a high-profile researcher, it was a spectacular, humiliating and public fall from grace, yet through gritted teeth, dogged determination, and the emotional and financial support of many loyal fans, Hwang has quietly rebuilt his career. He now works at a shiny new laboratory that he founded on the outskirts of Seoul – the Sooam Biotech Research Foundation – where he understandably shies away from human cell research, and instead focuses on cloning big animals.

Hwang has spent the last two decades making cloning look easy. In 2005, he produced the world's first-ever cloned dog, a goofy black-and-tan Afghan hound called Snuppy*. It took Hwang and his team two and a half years (or 19 dog-years) to produce the pooch, but since then he's honed his methods, knuckled down and cloned more than 500 dogs.† Many are copies of beloved pets, but he's also

---

*'Snuppy' is short for 'Seoul National University puppy', but with the animal now fully grown, perhaps his name should be changed to 'Seoul National University dog', or 'Snug' for short.

†Advice from Sooam's website: 'When your dog has passed away DO NOT place the cadaver inside the freezer. Then, patiently follow these steps: (1) Wrap the entire body with wet bathing towels. (2) Place it in the fridge (not the freezer) to keep it cool. Do not forget it is in there, or you'll have a nasty shock when you go for the milk … (OK, I added the last bit in, but it's a valuable piece of advice.) Please take into account that you have approximately five days to successfully extract and secure live cells.'

cloned several dozen working dogs for the Korean National Police Service, not to mention cows, pigs and coyotes. Who better, then, to bring back the woolly mammoth? In 2012, Sooam entered into an agreement with the keepers of the exquisitely preserved Malyi Lyakhovski mammoth at the NEFU's Mammoth Museum. Under the banner of the Mammoth Resurrection Project, the Koreans would supply their cloning expertise if the Russians handed over the best mammoth tissue they could find. 'It's a very exciting experience for us,' says Sooam scientist Insung Hwang (no relation to Woo Suk). 'Sooam have lots of experience cloning large animals and the Russians have a lot of mammoths. It's the perfect partnership.'

And so it was, in March 2014, after the Malyi Lyakhovski mammoth – or 'that bloody mammoth' as I like to call it – travelled all the way from its frozen resting-place to the NEFU in Yakutsk, that Hwang and an all-star cast of mammoth experts took part in one of the most bizarre spectacles of recent times – a CSI-style autopsy of an Ice Age mammoth. The dirty, slate grey carcass lay on an enormous slab, weirdly juxtaposed against the sterile, white-tiled laboratory, while the scientists, dressed in full forensic garb, jostled one another for samples. Under the watchful eye of two TV crews they prodded, poked, measured, photographed, drilled, sawed and plundered the slowly thawing carcass. No part of its anatomy was left untouched. If the scientists were to make the most of this remarkable animal, they would need to act quickly before decay set in.

They could tell from the curvature of the animal's tusks, its nipples and its internal plumbing that the Malyi Lyakhovski mammoth was female. From her tusks, scanned in 3D at the local hospital, they could tell also that she had lived a long and productive life. Mammoth tusks grew throughout their lives, leaving a series of growth rings from which it is possible to deduce not just their age but snippets of their life story. The tusks of adult females, for example, would have grown more slowly when the animal was pregnant or lactating. This mammoth, tests revealed, had given birth to eight calves; seven of these survived past weaning, and one died while still dependent on its mother's milk. Then, after 30 years of childcare, the pattern of growth rings changed again, most likely when the Malyi Lyakhovski mammoth rose to become the matriarch of her herd. After reaching the pinnacle of mammoth society, however, life became difficult. Mammoths, we know, had just four teeth – two in the upper jaw and two in the lower. With rows of ridges, these molars were great for grinding plants, but when the grooves wore down the teeth fell out. New ones would grow in their place, but when the sixth and final set came loose, it was game over for the gummy beast. When she died in her fifties, the Malyi Lyakhovski mammoth was down to her last few gnashers, themselves far from pristine. In her stomach and liver the scientists found a number of hard, round stones, which they think are either gallstones or actual stones that she swallowed accidentally. Then, one unseasonably warm day towards the end of the last Ice Age, she came across a boggy pool and leaned down to take what would be her last drink. Her forelimbs became stuck in the mud and no amount of struggle could free her. Teeth marks on her bones and the grizzly state of her gnarled upper body tell us that the Malyi Lyakhovski mammoth was eaten alive. A prolific mother and mammoth of great status, she met with a brutal and gory end.

But her story, of course, doesn't end there. 'It's amazing that her body is with us at all,' says Roy Weber from Denmark's Aarhus University, who attended the autopsy. 'She must have fallen into the bog on the one day in 40,000 years that it wasn't frozen. Then it stayed frozen ever since.' It's this continuous, uninterrupted, long period of freezing that means the mammoth is still with us today and the autopsy could happen at all. 'It looked like an animal that had died two or three weeks ago, not one that had been dead for tens of thousands of years,' he says.

When researchers cut into the mammoth's body, they found that parts of it were still remarkably fresh. 'The muscle looked like beef steak from the supermarket,' says Weber. And when samples of the apparently 'liquid blood' were tested, they were found to contain traces of haem.* The find hints that the fluid is perhaps some dilute, degraded version of mammoth blood, but why the 'blood' remained liquid when the rest of the mammoth froze remains a mystery until the liquid has been fully analysed. The most likely explanation, says Weber, is that the substance is a mix of sediments and breakdown products from blood and other tissues, which could have lowered the liquid's freezing point, and that natural anti-freeze molecules produced by bacteria inside the mammoth could have helped prevent the substance from turning solid.

For the South Koreans, however, it was all good news. Mature mammalian red blood cells don't contain nuclei so can't be used for cloning, but the seemingly fresh flesh raised hopes that cells suitable for cloning might be found. 'Other well-preserved mammoths have been found but it could be that the bleeding mammoth proves better at the cellular level,' says Insung Hwang. During the autopsy, Russian scientists examined tissue samples from the mammoth and were enthused to spot what look like muscle cells, but it's one thing to spot the outline of a cell on a microscope slide and quite another to find a cell whose DNA is sufficiently intact to be used for cloning. Another problem is the omnipresent Russian red tape; samples due for export are all too often delayed, meaning that precious time and potentially precious cells are lost long before they ever make it to the shiny, new labs at Sooam in South Korea.

With this in mind, Hwang and colleagues decided to move the mountain to Mohammed … or, at least, the lab to the mammoth. The South Koreans have funded the set-up of a purpose-built laboratory in Yakutsk itself, and because a woolly project deserves a woolly name they've

*A constituent of the oxygen-transporting haemoglobin molecule found in vertebrate red blood cells.

called their new facility the International Centre for the Collective Use of Molecular Palaeontology for the Study of Cells of Prehistoric Animals, or the ICCUMPSCPA for short. Russian scientists have gone to South Korea to learn how to clone large mammals, and the plan is that as fresh tissue is found it will be studied *in situ* on Russian soil, where researchers will have the know-how and kit to initiate a cloning attempt.

So let's just take a breather and recap. The race is on to bring back the woolly mammoth. Two different teams of scientists – one from South Korea, the other from Japan, both of which know a lot about cloning – are collaborating with two different Russian research institutions – who have lots of mammoths. The Japanese have ginger toddler Yuka, while the South Koreans have 'that bloody mammoth'. Both groups hope to extract nuclei from the dead mammoths' cells and then use them for cloning. Both are optimistic. They have to be. To take on a project of this vision and magnitude necessitates a 'glass half-full' disposition, but there are plenty of sceptics who think the projects are doomed to fail. The chances, they say, of finding a mammoth cell with a viable, intact nucleus are so remote as to be almost impossible. 'I have an antique porcelain vase on my mantelpiece,' says physiologist Kevin Campbell from the University of Manitoba in Winnipeg, Canada, who has studied mammoth molecules. 'It looks beautiful but I don't put water in it because it leaks. It's the same with mammoth cells.' Even a cell that looks good down the microscope is likely to be damaged and its contents long since leaked away. Just as well, then, that there's a third horse in this race.

Edit an Elephant

Harvard University's George Church is also trying to revive the woolly mammoth, and if anyone can do it, I believe he can. My conviction is based on two reasons. First, as I have discussed already, Church is one of the most intelligent, innovative, brilliant geneticists that there is. He has a string of high-profile scientific achievements under his belt and the most sophisticated gene-editing technology available at his fingertips. Second – and here's the deal breaker – he looks a lot like God. Church has a beard of biblical proportions, the likes of which wouldn't look out of place in a Michelangelo fresco. If I'm to put my trust in one man, it's got to be the one who would look most at home on the ceiling of the Sistine Chapel. His plan is to use genome editing to make something that, to all intents and purposes, will look and act like a woolly mammoth – a big, hairy elephant that loves the cold and can live in the Arctic.

In their time, woolly mammoths evolved many adaptations to cope with the extreme cold. On the outside, they were famously woolly. Their skin was pitted with numerous sebaceous glands that helped waterproof the fur by keeping it oiled. They had little ears and tails to minimise heat loss, and a spare tyre so thick it would make a sumo wrestler look dainty. But there were also changes on the inside. In 2010, Kevin Campbell and colleagues demonstrated that woolly mammoths carried an odd version of the haemoglobin gene. By inserting this mammoth gene into a bacterial cell, they persuaded the bacteria to make mammoth haemoglobin and were able to show that this Pleistocene molecule was particularly good at offloading oxygen at low temperatures. This would have helped mammoths to conserve energy and cope with the cold. 'It's genuine, authentic haemoglobin,' says Campbell. 'It's no different to what you'd get if you went back in time and took a blood sample from a real, living mammoth.' He hadn't quite de-extincted the mammoth, but he had de-extincted one of its proteins.

Then in 2015, two groups published high-quality versions of the mammoth genome. One group, led by Vincent Lynch from the University of Chicago, compared the woolly mammoth genome to that of the Asian elephant, the woolly mammoth's closest living relative, with which it shared a common ancestor just six million years ago. Although the vast majority of the nucleotides were identical, the researchers found 1.4 million DNA letters that were different and which in turn altered the sequence of more than 1,600 protein-coding genes. Some of these genes were involved in hair growth and colour, some in sensing temperature, while others helped regulate the internal body clock, a possible adaptation to life in a place where the summer sun sometimes never sets.

Together these genetic idiosyncrasies provide a road map for anyone seeking to resurrect the woolly mammoth. Much as a would-be Neanderthal-maker could theoretically take a human cell and edit it to become Neanderthal-like, Church plans to take an elephant cell and alter its genome so that it becomes mammoth-like. 'But just because there are millions of differences [between the mammoth and elephant genomes], it doesn't mean it's a daunting task,' he says. Far from it. To fulfil his goal of making a 'mammoth-like' creature, he doesn't need to edit all of the million or so uniquely mammoth sequences into an elephant cell. Instead, he plans to alter only the most salient of these genetic differences.

He's using a technique called CRISPR (pronounced 'crisper'). This new method enables genomes to be edited with pinpoint accuracy. In the old days, researchers could add genes into cells but there was no saying where in the genome the new additions would end up, raising fears they might disrupt cell growth-related sequences and cause cancer. CRISPR, in contrast, is seen as a safer option because it can be used to edit the genome at precise locations.

CRISPR stands for 'Clustered Regularly Interspaced Short Palindromic Repeats'. As the fiendishly fashioned name suggests, these DNA sequences are clustered together, regularly spaced and repeat in palindromic order; drab as a fool, aloof as a bard, some of the DNA letters read the same forwards as backwards. They're found naturally in some bacteria where they help fight off viruses. Co-opted for use in the lab, the system has been likened to a pair of molecular scissors being guided by a satnav. The satnav is a specially designed guide molecule made of the DNA-like molecule RNA, which directs the scissors, and an enzyme called Cas9, to snip the genome at the exact desired spot.

The technology was propelled into the spotlight in 2012, when Jennifer Doudna at the University of California, Berkeley, and Emmanuelle Charpentier, now at the Helmholtz Centre for Infection Research in Braunschweig, Germany, co-authored a paper showing how they could use the system to cut the double-stranded genome at any place they wanted. Then a year later, George Church and colleagues showed that the system could be used not just to cut but also to edit the genomes of human cells, a milestone in the CRISPR story. CRISPR, we now know, can be used to add, delete or alter anything from single nucleotides up to whole genes. With the technique being relatively cheap and easy to use, CRISPR has since been adopted by hundreds of laboratories all over the world, which are using it to help understand how genes influence health and disease, to design and develop new therapies, and to do lots of other clever stuff. But it's also been shown to work in other species … including elephants.

Under the banner of the 'Mammoth Restoration Project', Church and colleagues have already used CRISPR to edit the genes for mammoth haemoglobin into the nucleus of an elephant cell. Then, when he's satisfied that the elephant cell can read the new instructions and make *bona fide* mammoth haemoglobin, he'll turn his attention to other key mammoth traits, like fur and fat. To give his animal the required shaggy coat, he'll probably have to alter around half a dozen different genes; mammoth fur was, after all, anything but ordinary. Where the coarse, outer hairs on their flanks and belly grew up to a metre (3.3ft) long, the finer, curly under-wool was much more closely cropped.

He'll also be able to engineer coat colour. Mammoths, we know, carried a particular version of a gene called *Mc1r* that influences skin and hair colour. But depending on the exact sequence he chooses to use, Church could make hairy elephants that range in colour from strawberry blonde to deep auburn. Church tells me he is plumping for ginger. 'There would have been mammoths with light ginger hair,' he says. So why not?! And because he is aiming to recreate features that would help his animal to survive in the Arctic, rather than an exact genetic copy of the mammoth, he's not averse to the idea of adding in genes from other species too. One option would be to plagiarise DNA from another shaggy-coated Arctic mammal, the musk ox (*Ovibos moschatus*). 'Then there are woolly humans,' says Church. People with hypertrichosis or 'Werewolf Syndrome', for example, are completely covered in hair. Courted as circus sideshow freaks in times gone by, it's now known that some sufferers owe their fate to a particular genetic mutation. Cut and paste that into the elephant genome and it may be possible to generate one seriously hairy beast. As I write this now, Church has 'mammoth-ified' over a dozen different positions in the elephant genome, but that's the easy part. The next stage of the process, making an embryo that develops normally, is far more challenging.

An Elephant in the Room

The problem is that all of the various mammoth-making methods being discussed at some point hit the same two stumbling blocks. They all involve using elephants and elephant eggs, neither of which are easy to come by.

The first problem, then, is to find a female elephant that is ovulating. An adult female releases just one egg every four months, but a few weeks before this happens, she lets slip some pheromone-laden mucus from her vagina, which signals her fertility to pachyderms near and far. Close by, her immediate family amplify this sexual signal via the 'mating pandemonium'. Herd members charge around chaotically, trumpeting and screaming, advertising the female's readiness for sex to anyone that wants to know. It's a bit like Newcastle on a Saturday night. The resident bull, who may be tens of miles away, picks up on these not-so-subtle cues and rushes to be near her, so that they can mate as soon as she ovulates. The result: in the wild, at least, there are almost no fertile females that aren't already pregnant.

Best then to focus efforts on captive elephants, but even then there are problems. Female elephants endure the longest pregnancy of any mammal – 22 months – then spend roughly the same amount of time nursing their calves. So because mammals tend not to ovulate while they are pregnant or producing milk, this means it could be four or five years from one egg being released to another being available. This limits the potential pool of eggs for mammoth restoration.

But suppose an ovulating female is found. The next step would be to collect the egg, a procedure that involves navigating the complex system of internal plumbing that exists between an elephant's ovaries and the outside world. Where humans have but the smallest of gaps between daylight and the entrance to the vagina, elephants have an enormous, metre-long tube called the vestibule. The vestibule is not, as its name might suggest, a space for hanging up coats, nor do male elephants have to knock before they enter. It's simply a quirk of elephant biology – where humans have but the smallest of gaps between daylight and the entrance to the vagina, elephants have this additional waiting room. Bull elephants may have the largest penis of any land mammal (over a metre long when erect) but the vestibule is as close to the vagina as the wayward, bendy member ever gets. Because of this, an elephant's hymen, the membrane that covers the entrance to the vagina, isn't physically damaged during sex. Instead, sperm have to wriggle through a tiny hole in the hymen to reach the vagina and beyond, and when a baby elephant is born and the hymen is ruptured, it just grows back. To retrieve an egg, a would-be mammoth-maker would need to guide a surgical instrument to places an elephant penis has never been: up the vestibule, through the hymen, straight on at the vagina, to infinity and beyond. It's a tricky manoeuvre, but it has been done. A team of specialist veterinarians led by Thomas Hildebrandt from the Leibniz Institute for Zoo and Wildlife Research in Berlin (of whom much more in Chapter 8) has refined the technique, and managed to harvest elephant eggs.

The next step, then, is to use those eggs to make a mammoth. Embryonic development needs to be kick-started in a petri dish. But no one has ever made an elephant embryo in the lab, much less a mammoth one. Cloning, the method most often mooted, is notoriously inefficient. When researchers were trying to de-extinct the bucardo, they used over 500 goat eggs to make over 500 cloned embryos, most of which stalled in the culture dish. The best of the bunch – a couple of hundred – were then implanted into over 50 different surrogate goats, of which only seven became pregnant.

But if the odds weren't great for the bucardo, they'll be even worse for the woolly mammoth. The bucardo cloning experiments sought to reprogramme DNA from healthy adult skin cells, frozen carefully with protective chemicals. But if Ice Age tissue is to be used to make a mammoth, then the cells and any DNA in them will be far from perfect, lengthening the odds of success even further. Even if the DNA is decent, crafted perhaps through gene editing, there's no way of knowing if an elephant egg can reprogramme mammoth DNA. And if the bucardo researchers had to make hundreds of cloned embryos in order to generate one live birth, then a mammoth-maker would probably need to make thousands. The little bundles of dividing cells would then need to be transplanted into the womb of a surrogate, but here again we sail in unchartered territory.

No one has ever transplanted an embryo back into a female elephant. To do this, the embryo would need to be guided up the vestibule, through the hymen, straight on at the vagina, keep going at the cervix, then perform an emergency stop at the uterus. It's a torturous, twisty, two-metre (6.6ft) journey. The plucky vet would need either the arms of Mr Tickle or an eye-watering endoscope, but there could be another route. Sooam researchers plan to return their mammoth embryos not via the vestibule but up the rectum. A laparoscopic device would then be used to punch a hole in the intestinal wall and set the embryo down close to the womb. It's a reproductive shortcut, devised for rhinos, that shaves over a metre off the embryo's journey. If Option A is like taking the ring road, then Option B is like going cross country, through a field full of manure. Even then, the chances that the little embryo would survive are slim. No one knows how a mammoth or genetically modified elephant embryo will fair inside the uterus of an elephant. But we do know that when the embryo of one species is placed inside the uterus of another, things tend not to go well. Team Bucardo, remember, transplanted embryos into 50 different surrogates, of which only seven became pregnant, but six of those pregnancies failed and the one cloned bucardo that was born died in the arms of the vet who delivered her. Experience from the bucardo project, and from attempts to clone other endangered species, suggests that if researchers try to make mammoth embryos and have them develop inside the wombs of surrogate elephants, there will be many, many failures before even coming close to success. The surrogates will be placed under enormous stress, guinea pigs for experimental procedures that have yet to be tested.

And it's now that the very large elephant in the corner of the room, the one that's been sitting quietly throughout the whole of this chapter, stands up, waves a flag that says 'OVER HERE' and starts to loudly trumpet its displeasure.

Asian elephants are the woolly mammoth's closest living relatives, so it is to them we must turn for eggs and surrogacy skills. But Asian elephants are endangered. Their numbers have been reduced by half over just three generations, yet we continue to hunt them and trash their habitat. These are not laboratory animals. They are not experimental tools. A living, breathing mammoth might well be a marvellous thing, but at what cost?

Scientists are, of course, trying to think of ways around these problems. Maybe researchers could collect eggs from elephants that have died naturally or been culled. Maybe eggs could be created artificially in the laboratory (see Chapter 9), or clever tweaks to the cloning process could increase the chances of success. Perhaps, thinking much further ahead, one day we could do away with the need for surrogate animals altogether and grow embryos in the lab in some sort of artificial womb. But if, as the old saying goes, ifs and ands were pots and pans, there'd be no work for tinkers' hands.

Woolly Thinking

But suppose for a minute that someone does manage to make a mammoth. What happens next is the sad part. If – and it's a big if – a healthy baby mammoth is born, it runs the risk of being very lonely indeed. We can surmise from mass mammoth graves and from fossilised footprints that mammoths once lived in large, extended families. Like modern-day elephants, these groups were probably matriarch-led, and largely female. Herd members would have communicated with one another via low frequency noises and it's reasonable to assume that, just like elephants, woolly mammoth mothers would have had close relationships with their calves. But would a woolly mammoth calf recognise its elephant mum? Would it know to suckle?

How about the mother? Would a modern elephant mum caress her hairy newborn with her tactile trunk, or reject it as an aberration? And if she did decide to care for a Pleistocene misfit, how could she possibly teach it to act like a mammoth rather than an elephant? If the goal is to repopulate Siberia with mammoths, how would the twenty-first-century lookalikes learn to survive in the cold when their female role models are sun-loving? The degree to which mammoth behaviour is innate and/or learned is a very woolly area. With so much at stake, it's inevitable the first de-extincted woolly mammoth would be born and raised in captivity where the environment could be carefully controlled. But then other woolly mammoths would follow. They'd have to. It's not just a mix of genders that is required, but a mixture of genes. Without genetic diversity to buffer against disease and environmental change, the new animals wouldn't stand a chance. If the new mammoths were to hail from just one or two genetic founders, over time they'd end up as inbred as the royal family, albeit arguably with smaller ears. The goal, then, would be to release herds of mammoths back into the wild. But how many to release and where? Here again, predictions are vague.

By matching the chemical isotopes in tooth enamel with those in the soil, researchers calculate that some mammoth species roamed up to 300 miles a year, so they would need a lot of space. Modern elephants eat up to 200kg (450lb) of food per day, so the mammoths would also need a lot of fodder. If the aim is to restock Siberia with woolly mammoths then we have to question whether the reserves being prepared for them are big enough. Pleistocene Park, the brainchild of Sergey Zimov, covers around 160km^2 (60 square miles) while a second, more southerly nature reserve created closer to Moscow is 50 times smaller. Sure they could roam elsewhere. Not many people live in the High Arctic. But if a key goal is to keep mammoths in such numbers that they can help keep the

Arctic frozen and its sequestered carbon tucked away, then it's time for some joined-up thinking. Even if all the technical hurdles involved in making a mammoth were overcome tomorrow (which they won't be), it would still take well over half a century to generate anything like a single viable herd … by which time, if current predictions are to be believed, the Arctic ice could already be gone. 'The mammoths would arrive too late to do any good,' says climate engineer Hugh Hunt from Cambridge University. With no ice to reflect the sunlight back into space, the world will continue to warm. The tundra will melt and the carbon time bomb that it's holding will explode. 'We need to be looking at refreezing the North Pole in the next 10 to 20 years,' he says, 'after that it will be too late.' If we're looking to the woolly mammoth to save us from global warming we need to think again. Although the image of a mammoth in pants and a cape is visually appealing, mammoths are simply not the climate-change superheroes that some would paint them to be.

Were I to choose one animal to bring back to life, would it be the woolly mammoth? Sure, I'd like to see one. I'd cough up the cash for a thermal onesie, a brandy-filled hip flask and an airfare to Siberia in order to marvel at the wonders of a fully stocked Pleistocene Park, but only if the mammoths could be made without harming the pachyderms we already have. I don't want mammoths back at the expense of the elephant. I don't want high-risk, potentially harmful procedures practised on a species so loved, so precious, so very much in need of our protection. There will, perhaps, come a time when science and technological progress will triumph over my objections, in which case, so much the better. But for now, at least, the woolly mammoth is off my list.

CHAPTER FOUR

King of the Birds

As the lid is slowly lifted, I'm not sure what to expect. 'There it is,' says the museum curator with a wistful smile. 'It looks rather sad, doesn't it?' I have to agree. I'm looking down at the most famous dodo in the world, but all there is is a desiccated head. It lies on its side in a cardboard box, its one visible eye half closed. Its wizened face, with large protruding beak, is framed by a taut balaclava of black, leathery skin with a few stubby feathers sticking out. Next to it is what looks like the bird's mirror image, another dodo staring back at the first through a single hollow eyehole. But it is, in fact, part of the same bird. When Victorian scientists scrutinised the dodo, they peeled away the skin from one half of the skull then laid it out like some bizarre death mask. In another, much smaller box there's an ossified eye socket. In a third, a bony foot and some bits of dried, scaly skin. It looks like a post-pub binge KFC.

These, then, are the remaining body parts of the celebrated Oxford dodo, kept with care and respect at the university's Museum of Natural History. So unusual is the exhibit, it inspired Oxford University don Charles Dodgson (aka Lewis Carroll) to incorporate the dodo into the children's classic *Alice's Adventures in Wonderland*. But just as the Cheshire Cat disappeared until all that remained was a grin, so too the Oxford dodo, a resplendently full specimen, disintegrated ... until all that was left was this. It may 'only' be a head and a foot, but with dried skin and flesh, it is the best-preserved dodo in the whole wide world. Other museums may have dodo bones, but there really is nothing quite like this.

This is the story of how this remarkable bird made a journey spanning thousands of miles and hundreds of years to arrive in Oxford where its DNA was decoded, and how it inspires, in me at least, the dream of bringing the dodo back to life. I was expecting something gruesome and macabre, but what I find is something sad and beautiful. This bird, desiccated and dissected, ancient and iconic, carries with it an unexpected air of dignity and calm. This is the face of the last dodo on Earth.

As Dead As

As dead goes, it doesn't get any deader than a dodo. The poster child of extinction, they even get their own idiom. To be as dead as one, the saying goes, is to be unambiguously and unequivocally not alive. The phrase implies there are degrees of dead-ness; that some things are deader than others. If there were a scale of 'deadness', dodos would be off it. Dodos are the deadest things there are. So if ever there were a candidate for de-extinction, surely the dodo would have to be it? To be 'as alive as a dodo' may not have quite the same ring, but it would certainly make a nice change, not least for the dodo.

The dodo was a large flightless pigeon that once lived on Mauritius, a small island in the Indian Ocean thousands of miles from the East African coast. Yet despite its prominence in our collective consciousness, we actually know very little about it. What we do know comes from the study of its remains – a handful of mummified body parts, thousands of disarticulated bones – and from the reports of those who saw or heard about this remarkable creature. Seventeenth-century sailors who visited the island painted and sketched the bird, and wrote descriptions of it in their journals, but their colourful reports are full of inconsistency and contradiction. According to the various records that exist, the birds were easy to catch, they were difficult to catch. They were slow, they were agile, they were smart and they were dumb. Then, as the sailors returned home and news of the dodo spread, other artists, most of whom had never seen the bird alive, also began to paint it. They used artistic licence to fill in the gaps in their knowledge and satisfy their patrons' preconceptions. As a result the dodo has been variously portrayed as fat, thin, stooped, straight, bumbling, athletic, pigeon-toed, web-footed, brown, grey, black and blue. Like some paint-fuelled version of Chinese whispers, the birds became less like their true selves and more like the big-bummed bungling caricatures many of us think of today when we imagine the dodo.

With their oddly proportioned bodies and comical looks, they were all too easy to poke fun at. Dutch seafarers dubbed them '*dodaersen*', or 'fat-arses'. It's an insult that may have morphed into the name we know them by today, and that may also have laid the foundations for a major insecurity complex. Was the dodo the first animal to fret about the size of its behind? Did it wander the forests of Mauritius lamenting its puffy plumage and pondering that apocryphal question: 'Does my bum look big in this?' Even the great wig-wearing eighteenth-century scientist Carl Linnaeus joined in. According to the dodo's reputation, Linnaeus bestowed on it the scientific name *Didus ineptus*.

And although the official moniker has since changed,* the bird's common name is still used in derogation. Ask most people what they know of the dodo and they'll tell you they were stupid, fat and sluggish. An evolutionary disaster, they were too dim to sidestep the blows of those who came to kill them and too slow to outpace extinction. The truth is, it's all a bit unfair. It's too easy to poke fun at something we know little about, which is no longer here to defend itself or otherwise prove us wrong.

What I can tell you is this. Some time in the last seven million years (no one is really sure when), the dodo's ancestor – a much smaller, airborne pigeon – landed on Mauritius. The island paradise was a welcome stopover, so much so that the birds decided to stay. With no natural predators and a forest floor littered with fallen fruit, the birds found they did a lot less flying and a lot more walking. Flight, after all, requires a lot of energy, so it's far easier to not bother if you don't have to. Then, sculpted by evolution over many generations, their wings began to shrink until eventually they lost the ability to fly altogether. And they got bigger. According to the 'island rule', island-dwelling species change in size depending on the resources they encounter. Bigger mammals, delightfully, shrink; there was a time when there were mini mammoths on Cyprus (*Mammuthus creticus*) and hobbit-sized humans (*Homo floresiensis*) in Indonesia. Rodents and birds, however, tend to grow bigger. Minorca had its giant dormouse (*Hypnomys mahonensis*), Madagascar boasted the enormous elephant bird (family Aepyornithidae) and Mauritius witnessed the rise of the dodo.

Although no one knows exactly when the dodo evolved, nor how many there were during the bird's heyday, we do know that 400 years ago, life was good for the dodo. The giant pigeons lived amidst dense forests of ebony and palm, shooting the breeze with exotic birds and giant tortoises. They scrumped for fallen fruit, built their nests

*It's *Raphus cucullatus* nowadays, since you ask.

on the ground and raised their young unbothered by any menace.

But a distant speck on the horizon was about to change all that. In September 1598, Dutch ships *en route* to the East Indies spied the island idyll from afar. The sailors, who had been at sea for months, were exhausted, hungry and out of fresh water. So they weighed anchor offshore and rowed towards the beach. What they found was like manna from heaven – birds so plentiful they could swipe them from the sky with sticks, fish so abundant they were netted with ease and giant tortoises so compliant they could (as one early picture shows) ride them along the beach. And then there was the dodo.

'They walked upright on their feet as though they were a human being,' said one record from the time. It had 'the body of an ostrich', 'three or four black quills' instead of wings and a 'round rump with two or three curled feathers on it'. 'Their war weapon was their mouth, with which they could bite fiercely,' said another report. So when the sailors walked up to them and clubbed them to death, it wasn't that they couldn't fight back, it's that they chose not to. Their problem was not one of weaponry, but of attitude. 'Modern pigeons are uniformly aggressive across the board,' says present-day ornithologist and dodo expert Julian Hume from London's Natural History Museum. 'Dodos were unlikely to be any different. But these birds had never seen humans before so they didn't perceive them as a threat.' Dodos, it seems, were suicidally curious and hopelessly trusting of humans. The cry of a captive dodo would draw others from the forest, which could then be grabbed as well. It's because of this that people thought them stupid. 'But they weren't,' says Hume, 'they were just very naïve.'

Faced with an exotic, unique giant of a bird that was like nothing they had ever seen before, the hungry sailors thought of one thing only – what it would taste like. So they killed them and carried them back to the ships' kitchens. What a disappointment that was. Unlike most exotic animals, the dodo didn't taste 'a bit like chicken'; it was, apparently, 'offensive and of no nourishment'.* 'Although we stewed them for a very long time,' one sailor wrote, 'they were very tough to eat,' so the dodo received another nickname: '*walghvogel*', or 'repulsive bird'.

For those dodos unlucky enough to be captured and cooked, it was an ignominious death, but for those left alive on the island it was the start of a much slower passing. Ideally situated as a stopover for their fleets as they criss-crossed the Indian Ocean, the Dutch visited Mauritius many times before setting up a permanent base there in the 1630s. Along the way, they wrecked the dodo's natural habitat, felling forests to make way for sugar plantations, and flooded the island with invasive species. Rats, monkeys, pigs and goats competed with the dodo for resources. They destroyed the dodos' nests and predated their eggs and chicks. The dodo went into steep decline.

Back to Blighty

Some birds, however, left the island alive. Sailors recognised there was money to be made from unusual-looking creatures if they could sell them to collectors on faraway shores. So the Oxford dodo was bundled into a crate and shipped to Great Britain. Cooped up in a crate for months on end, it must have been a miserable journey, but it's thought that the Oxford dodo survived and made it to London alive. English theologian Hamon l'Estrange describes an encounter with what is presumed to be the Oxford dodo in 1638. He was walking through the back streets of London when he stumbled across a board outside a house advertising a 'strange looking fowle'. Inside, up some stairs, he found a stocky, turkey-sized bird kept in a cage. The keeper fed it pebbles – 'some as big as nutmegs' – to aid its digestion, and

*So more like a doner kebab.

called it a dodo. Here was proof that one dodo at least had made it to Europe alive. But when the bird fell off its perch, as it inevitably did, the gawping didn't stop.

First it ended up in a pay-to-view collection in South London called the Tradescant Ark, which listed in its catalogue a 'dodar, from the island Mauritius it is not able to flie being so big.' But when the owners of the collection died, the dodo passed into the hands of family friend Elias Ashmole, who put it on display in Great Britain's first public museum, the newly built Ashmolean at Oxford.*

The Oxford dodo was finally in Oxford, but its future was not looking rosy. Back then, museums were fairly hands-on affairs, with visitors encouraged to pick up and manhandle the exhibits. With its oddly shaped proportions and unusual beak, the dodo must have seemed particularly inviting. Little by little, passed from pillar to post, the Oxford dodo began to deteriorate. Insects invaded its poorly preserved body and in 1755, when the museum's trustees met for a routine inspection of the exhibits, they decided that the Oxford dodo was so ravaged it should be destroyed and replaced with a better dead dodo.

But there was just one problem. There were no more dodos to be had. In Mauritius, the last of the dodos was gone by the end of the 1680s. Extrapolating from this date, present-day experts estimate that the species became 'as dead as' sometime around 1693. With all living dodos gone, the Oxford dodo was irreplaceable. What happened next has become conflated into Oxford folklore. Unaware that the dodo had become extinct, museum staff did as they were told and chucked the Oxford dodo onto a bonfire where its body quickly burst into flame. But then, at the eleventh hour, an asbestos-fingered museum assistant

*In its early days, the Ashmolean met with mixed reviews. With its seemingly random mish-mosh of trivia, critics called it the 'Knickknackatory', which did not go down well with Ashmole. Apparently it was all too much and he 'relapsed into the Gout'.

decided to defy protocol and pluck what was left of the bird from the flames. A chargrilled head and a single foot were retrieved and returned to the museum, where I visited them more than 250 years later.

It's a story that infuriates the dodo's current protector, Collections Manager at Oxford University Museum, Malgosia Nowak-Kemp. The truth, she explains to me on the day I make my own inspection, is far less dramatic. 'There was no fire,' she says. The non-bony parts of the bird simply rotted away and were discarded. The conflagration myth emerged from a mistranslation – a Latin word for 'inspection' was mistaken for the Latin word meaning 'purification by fire'. 'The head and foot were put aside because they were the only parts of the bird worth saving.'

By the nineteenth century, the only known dodo remains were the Oxford head and foot, another foot in London, a skull in Copenhagen and a leg and bit of beak in Prague. So scant were the remains, so distant were the memories, some doubted the dodo had ever existed at all. The alternative, that it had existed then gone extinct, was unthinkable to the vast majority of Bible-believing literalists. Said one author of the time: 'The dodo … appears to have existed only in the imagination … or the species has been utterly extirpated … which is scarcely possible.' The Oxford specimen was, he said, an unknown species of bird that was still alive somewhere. For the Oxford dodo, it was a period of unprecedented existential angst. Had it ever existed at all?

Stuck in the Ashmolean with literally no body to turn to, the remains of the Oxford dodo idled the decades quietly. Then in 1847, came the first of two dissections. Victorian scientists teased apart the bird's head. Their goal, according to Nowak-Kemp, reached beyond intellectual curiosity. In a university then bereft of any science at all, the act served to raise the profile not just of the dodo but also of science itself. The scientists concluded that the dodo was not, as some had thought, some sort of a vulture or albatross, but a pigeon. The dodo was back in the game. People believed in it again.

The university built a new museum to house its scientific collections and to teach and research the sciences. In 1860, the Oxford dodo left the Ashmolean and moved across town to the new, cathedral-like Museum of Natural History, where it can still be found today. The museum is a Gothic masterpiece. If you ever find yourself in Oxford you must go and visit. On a sun-kissed autumn morning, its vaulted glass ceiling bathes the museum with light and warmth, illuminating the remarkable creatures presented there. It was with this in mind that just before the new millennium rolled in, Malgosia Nowak-Kemp quietly removed the Oxford dodo from public display and found it a new resting place, away from the bleaching rays of the sun in a cool, windowless backroom. The dodo that you see on display in the public gallery is a model, the skull next to it, a replica. But ask her nicely, and Nowak-Kemp, known also as 'the dodo lady', might show you the real deal. She has been looking after the Oxford dodo for the last 25 years and finally, it seems, the bird is being given the protection it has so desperately needed. The remains rest on tissue paper in specially selected storage boxes, acid free so as to minimise discolouration and decay. From time to time the boxes are moved as a precaution against theft. But when I visit, they are stashed in an elegant cabinet of curiosities amidst an eclectic collection of strange skulls, carapaces and horns. Nowak-Kemp handles the boxes with care and respect, laying them down on a nearby table as gently as a mother would a newborn. When the lid is lifted, we stare at the bird together. 'Do you ever get tired of looking at it?' I ask her. 'Never,' she replies.

We look but don't touch. I photograph but don't flash. With half the skull's skin removed, the scars of the Victorian dissection are obvious. But those of the second are far less visible. In 2001, the University of Oxford was home to a thriving ancient DNA research laboratory under the leadership of Alan Cooper, now at the University of Adelaide, Australia. For those interested in ancient DNA, it was a time of immense optimism. Researchers had extracted the molecule from Egyptian mummies, mammoths and Neanderthals, and Cooper himself had extracted DNA from another extinct and flightless bird, New Zealand's giant moa (*Dinornis giganteus*). 'In the early days of ancient DNA it was all about who could get their hands on the coolest specimen,' says ancient DNA researcher Beth Shapiro (now at the University of California, Santa Cruz), who joined Cooper in his lab round about that time, 'and it didn't get any cooler than the dodo.' They wanted to try to retrieve DNA from the Oxford dodo, not so they could resurrect the bird, but so they could use any DNA still left in the bird's shrivelled remains to help understand where the unusual bird perched on the tree of life.

The duo was granted access to the iconic artefacts. They began by retrieving samples of soft tissue, or 'crusties', as Shapiro calls them, by scraping away at the inside of the dodo's skull. But the samples yielded no DNA. So next they took a drill and bored a tiny hole into the dodo's leg bone. This time they hit the jackpot. The team was able to retrieve short snippets of mitochondrial DNA, from which they were able to confirm that the dodo was indeed a pigeon, whose closest living relative is the sultry, iridescent Nicobar pigeon (*Caloenas nicobarica*). Google it – it's a peach!

That such information could be gleaned from mitochondrial DNA was impressive, but the duo knew their limits. Any nuclear DNA that might exist would be present in such small amounts it would be impossible to detect with the techniques of the time. And with the Oxford dodo a precious one-of-a-kind, removing any further samples for analysis was unthinkable. So here, for now, ends the story of the Oxford dodo. Its remains under the watchful eye of Malgosia Nowak-Kemp, and long may it rest, preserved and peaceful, in the place that has offered it respite for the last century and a half.

But the search for dodo DNA goes on. The vast majority of all dodo bones ever found originate from a single source, an ancient dried-up watering hole located close to the coast of southeastern Mauritius. A little over 4,000 years ago, all sorts of animals, including dodos and giant tortoises, regularly visited the site, but then drought set in and the lake began to dry up. Animals continued to crowd around the dwindling waters, their excrement mixing with the increasingly salty and muddy mire. This, analyses of sediment cores reveal, helped to promote the growth of potentially toxic bacteria. Hundreds of thousands of animals became stuck in the mud and died through intoxication, dehydration and trampling.

Today the area, known as the Mare aux Songes, is a grassy swamp not far from the island's Sir Seewoosagur Ramgoolam International Airport, and the anaerobic sludge that killed so many animals in the past subsequently helped to preserve their remains. The fossils found at the Mare aux Songes are so common and well-preserved that the area has been designated a *lagerstätte* (German for 'storage space'), a term used to designate only the most exceptional of fossil sites. Recent excavations, by Beth Shapiro, Julian Hume, and others, have yielded hundreds of dodo bones. When the hapless birds became stuck in the mud 4,000 years ago, their top halves rotted away, but their mud-encased lower limbs were preserved. Many of the dodo bones found are leg bones.

Using cutting-edge techniques, Shapiro has tried to retrieve DNA from more than 50 different dodo bones pulled from the Mare aux Songes, but, aside from a few short mitochondrial sequences, she's had little joy. Although the bones are in good condition, the DNA inside their cells is not. A few thousand 'letters' or 'base pairs' of mitochondrial DNA is one thing, but it's a far cry from the billion or more base pairs of nuclear DNA that make up the complete genetic code of a bird. The dodo's nuclear DNA is either so badly degraded that there's genuinely none left any more, or the little there is is in such bad shape it remains beyond the reach of current methods.

I ask Shapiro if there's any small glint of hope, any conceivable chance of ever getting DNA from the dodo. She gives it to me straight. 'It's unlikely that we're ever going to find any specimens on Mauritius with DNA in decent shape.' The island is too warm, which makes DNA disintegrate. And the Mare aux Songes, where 99 per cent of all the dodo remains ever found come from, is too wet and too acidic, which makes DNA disintegrate. 'It's everything that you don't want when you're trying to preserve DNA.' The only hope of finding dodo DNA, says Shapiro, would be to find an expat dodo that died abroad and was then kept carefully at a cool, constant temperature over the centuries. It's not going to happen. We humans have a long history of *not* looking after the dodo – of killing it, eating it, destroying its habitat and then hanging around doing nothing while its ill-preserved remains fall to bits. If the most complete specimen we have is the Oxford dodo with its wizened head and time-ravaged snippets of mitochondrial DNA, then the bird's full genetic code is likely to remain forever beyond our reach. Without its genome there is simply no hope of bringing back the dodo. My dream of de-extincting it is 'as dead as … '. It's time to face the truth. I will never have the pet dodo of my dreams.

I'm gutted. Who wouldn't want to see a dodo for real? I'd dress up as Alice, give it a cane, put it in cufflinks and have it stage a Caucus race. That way, everyone would be a winner.*
As birds go, the dodo was one extraordinary pigeon, the likes of which we'll simply never see again. But the dream

*This is a joke for hardcore Lewis Carroll fans. Are there any? Thought not. At the risk of over-explaining a joke and sucking the joy from it … in *Alice's Adventures in Wonderland* the dodo organises a Caucus race in which participants can run in whatever pattern they like, starting and finishing at points of their own choosing, so everyone wins.

of resurrecting an extinct funky pigeon doesn't have to end there. There is another extinct pigeon, one that died more recently, that scientists think they *can* bring back.

An Avian Eclipse

When the dodos were dying out on Mauritius, a much smaller pigeon with wings that still worked was darkening the skies of North America. They weren't dumpy like the dodo, or sultry like the Nicobar pigeon – these were speedy and svelte. Males were cobalt blue with a peach-coloured breast, bright red eyes, and coral legs and feet. Females were, as females are, a little more subtle. At the start of the nineteenth century, the birds existed in mind-boggling numbers. There were billions of them. Their flocks were so large they literally blotted out the sun – an avian eclipse. Sometimes a flock took days to pass overhead. The collective beating of their wings was like the rumble of thunder and, some said, created a draught so powerful it actually chilled those on the ground below. Where the dodo, hidden away on its rural island idyll, had been easy to overlook, these birds, by their sheer numbers, were impossible to ignore.

The birds were passenger pigeons (*Ectopistes migratorius*), also known as the 'Blue Meteor', once the most abundant bird in North America, and probably the world. At one time there were more passenger pigeons than there were people alive on Earth. It's almost impossible to imagine, apocryphal even. But try … A single flock could be over 100 miles long and a mile across. If all the birds in that flock had flown single file, beak to tail, the line would have circled the Earth 22 times. If you took all the pigeons in the UK, multiplied them 400 times, then launched them into the air, *that* would be the size of a single passenger pigeon flock. And then, one day, they were all gone. The passenger pigeon went the way of its Mauritian cousin.

That the passenger pigeon could go from such enormous numbers to nothing in the space of a few decades is devastating, but we have no one to blame but ourselves. In their heyday, these immense flocks roamed the deciduous forests of the eastern and midwestern United States and Canada, gorging themselves on the acorns and beechnuts they found there. Like feathered locusts, they would appear *en masse* and ransack entire forests. Then, with their food supply exhausted, they'd simply take to the skies and move on. A seemingly inexhaustible supply of free protein that was all too easy to catch, we massacred them in their millions. So dense were their flocks that a single shot could slay dozens of birds at a time. There was no need to aim. A blind man could have pointed his rifle into the air and still brought them down. We clubbed them out of the sky with sticks and we blasted them with cannons. We poisoned them with whisky-drenched corn, set fire to their roosts and asphyxiated them with burning sulphur. We lured them into vast nets via the cries of captive pigeons that we tied to stools with their eyelids sewn shut, giving rise to the term 'stool pigeon' and, some time later, a song by Kid Creole and the Coconuts.

In the late 1800s, as the telegraph and railways spread, hunters found that they could hear about and travel to new nest sites all too easily. Passenger pigeons, the cheapest source of protein in the United States, became big business and the pigeon-ocide continued on an industrial scale. Bunged into barrels, the dead birds were sold by the tonne. Boasting a dark meat that was apparently delicious, most ended up baked, stewed or wearing pie-crust lids. Some people thought to stuff their pillows with the birds' feathers in the superstitious belief it would bring long life. It didn't. Not to the passenger pigeon anyway. Such was our greed that sometimes barrel-loads of unsold birds were left to rot. What a shameful waste. And if killing them wasn't enough, we systematically destroyed the forests they relied on for food and shelter. By the 1870s, European settlers had felled half of North America's native woodland for timber and agricultural purposes. It's hard to underestimate the effect this would have had on the birds that remained. With their nest sites dwindling, and the acorns and nuts that sustained them in short supply, their numbers began to plummet.

By the time anyone noticed and meaningful legislation was introduced, there were no passenger pigeons left to save. By the early 1900s, the few that remained were confined to US aviaries. Soon only one bird remained, a female called Martha, who lived in the Cincinnati Zoological Gardens, Ohio. Aged and immobile, visitors pelted her with sand to make her move, so keepers had to cordon off her cage. Then on 14 September 1914, Martha, the last passenger pigeon on Earth, succumbed to the inevitable. She died that afternoon and, some said, the United States' heart was broken. The passenger pigeon became nothing more than a memory.

Realising her significance, zookeepers froze her in a giant ice cube and shipped her to Washington, DC, where she was thawed, skinned, stuffed and put on public display at the Smithsonian Institution. She has since left her resting place only twice, once to appear at a San Diego conference, and once to visit her old haunt, the Cincinnati zoo. Both times, she was flown first class under the private supervision of a dedicated flight attendant. Born, raised and died in captivity, it's ironic that Martha flew further in death than she ever did in her 27-year-long life.

And that's the end of the passenger pigeon … except that it might not be. In 2012, a group of ornithologists, geneticists and conservationists, including Beth Shapiro and George Church, got together at Harvard Medical School to discuss whether or not this iconic species could be brought back to life. The meeting was organised by Ryan Phelan and Stewart Brand, who had been thinking about de-extinction and wondering whether it was possible. The experts concluded that it was.

Shapiro had already managed to tease nuclear DNA from the toe pads of museum birds, and gene editing technology was improving all the time. Sure there were technical hurdles, but the mood was optimistic. 'The meeting was a green light for us,' says Phelan. The duo went on to set up Revive and Restore, an influential, friendly, non-profit organisation dedicated to advancing the science of de-extinction (and more besides; see Chapter 8). 'The Great Passenger Pigeon Comeback' was hatched.

Today, the project is in full swing, at its helm a young scientist by the name of Ben Novak. Funded by Revive and Restore, and working under the guidance of Shapiro in her Santa Cruz laboratory, Novak is doing something that all of us want, but few of us ever manage – living his childhood dream.* He has been thinking about de-extinction since he was 13 years old. As a kid, he did a science fair project about the possibility of bringing dodos back to life. Then, a few years later, he opened a book and saw a picture of the passenger pigeon. 'I fell in love with the photo,' he says. 'I fell in love with the story.' He then began to eat pigeon,† sleep pigeon and breathe pigeon. So desperate was he to get things moving that a few years before he landed his job with Revive and Restore, he raised $4,000 from family and friends so that he could start sequencing passenger pigeon DNA on his own. He has researched their genetics, their history and their ecology, all with, what seemed at the time, a far-fetched dream of bringing them back to life, of giving their sad story a happy ending. So when the Revive and Restore project materialised, he jumped at the chance to get involved. 'He's smart. He has a can do attitude, and he has a lifetime of passion behind him,' says Phelan. He's also not at all what one would expect of a pigeon fanatic or an academic. Google 'pigeon fancier' and you'll find images of flat-capped pensioners smiling gummily for the camera as they

*My childhood dream was to open a hairdressing salon for long-haired guinea pigs. Sadly it never became a career option.
†Not literally.

coo over their prize birds. But not Ben Novak. He may have a flat cap but he wears his back to front. He looks like he should be in an indie rock outfit called 'Ectopistes' or a UK street dance troupe called 'Genetic Diversity'. A modest man, he has the nous and technical expertise of a post-doc 10 years his senior. Novak knows more about pigeons than the collective membership of the National Pigeon Association. If anyone can bring the passenger pigeon back, Ben Novak can.

His starting material is the hundreds of skinned, stuffed passenger pigeons that can be found in public museums around the world. Alongside Shapiro, Novak has extracted DNA from more than 80 such birds dating from the late 1800s to as far back as 4,000 years ago. As is to be expected for dead creatures of that age, the genetic material is broken into many, many tiny pieces. So Novak is using Next Generation Sequencing to read the sequences of all of the snippets in a sample in one go. By repeating the process over and over again, he can build up an increasingly accurate picture of the genome, albeit it in millions of tiny pieces. To date, Novak has sequenced 21 trillion base pairs of DNA. That's a staggeringly big number: it's '21' followed by 12 zeros.

The next step, then, is to reassemble these fragmented sequences into a full genome, first digitally inside a computer, and then functionally inside a cell. Help is at hand from the passenger pigeon's closest living relative, a rather average-looking coo-er called the band-tailed pigeon (*Patagioenas fasciata*), which has a very similar genetic sequence to the passenger pigeon. Just as researchers used the genome of modern humans to help thrash out that of Neanderthals, and the genetic sequence of the elephant has been used to help recreate the sequence of the woolly mammoth, Novak is using the band-tailed pigeon genome as a template against which to match and order the fragments of his passenger pigeon DNA. He now has the full genome from two birds, and bits and pieces of genomes from 37 others.

Next, he will use CRISPR, the gene editing technology being used by Church to mammoth-ify his elephant cells (see Chapter 3), to edit key passenger pigeon genes into band-tailed pigeon cells. Novak will effectively line up the virtual genomes of the passenger and band-tailed pigeons side by side and look for differences between the two. Then comes the tricky bit: working out which of the expected 10 million or so differences are really important. Which sequences, for example, gave the passenger pigeon its broad wings, its muscled chest and its long tail? How about its dappled cobalt wings or its rosy breast? Is there a genetic element to its wanderlust or its preference to flock in the billions? It's a daunting task.

The trick, says Novak, is to look for genes that have already been linked to certain characteristics, such as plumage and colouring, and then to tease out other, lesser-known genetic sequences that have been positively selected for. These are bits of sequence that have changed so quickly over time that they must have been shaped by something other than random genetic drift. They are stretches of the genetic code so useful that the birds that carried them were more likely to survive, reproduce and pass their genes on to future generations. Perhaps the sequences somehow altered wing shape or visual acuity, making the birds faster fliers or better foragers. Or perhaps they influenced the way their brains were wired, helping them to navigate vast distances. Positively selected sequences leave behind telltale signs in the genetic code that can be spotted by vigilant geneticists. It's these most salient of sequences that Novak will edit into his band-tailed pigeon cells.

Good! So we have, in theory, a passenger-esque genome inside a living bird cell. The next step is to make a baby bird. 'Why not do cloning?' I hear you shout. After all, why wouldn't you? You have read the previous chapters with intense, page-turning fury. You know about Dolly the sheep, Celia the bucardo and the world's would-be mammoth cloners. So could cloning be an option?

In mammalian cloning, donor DNA is injected into an egg, which, after a bit of biological jiggery pokery, turns into an embryo, which is then implanted into a surrogate womb. But with birds it's more complicated. The female's reproductive system is a bit like an automated assembly line. At one end, the ovary releases a fragile membrane-bound blob that contains the yolk and a tiny spot of DNA. Then, as the blob begins to tumble down the bird's oviduct, the white and then the shell are added in turn. By the time an egg is laid it is too developed, not to mention too covered in shell, to be of use for cloning, and even if it were, there would be no womb to return it to. Researchers have been trying and failing to clone birds for a decade or more, leading some to speculate that the procedure is simply impossible. 'The issue with cloning birds isn't just that it requires technical finesse, it is also a logistically complex process,' says reproductive biologist Michael Kjelland from Conservation, Genetics and Biotech, LLC.

Kjelland and colleagues have, however, been working on overcoming these hurdles. His idea is to get in early and collect the eggs as soon as they have been ovulated, but before the white and shell have been added. Working with chickens, because they are easy to get hold of, look after and produce lots of eggs, he uses a steady hand and a tablespoon to scoop out the grape-sized blobs. Next he uses a specially developed technique that combines microscopy with a chemical stain to visualise the egg's DNA. If cloning is ever to work, the egg's own DNA must be removed so that the donor DNA can be added, but locating it against the opaque background of the yolk has proved problematic. 'It was quite a stumbling block,' he says. With the yolky blob wedged inside a glass beaker, he then punctures the structure and uses a fine glass pipette to remove the DNA. 'The worry was that the yolk might spill out when we punctured it,' he says, 'but it didn't.'

The next step would be to add the DNA from some would-be cloned bird into the 'empty' egg and then coax the reconfigured cell to start dividing. But the developing egg would then need to be physically placed somewhere to be incubated. Here, Kjelland explains, there are two options. One is to transplant the embryo back into the reproductive tract of an adult bird, from whence it could tumble down the oviduct just like any other egg and have its white and shell added. The other is to carefully slice the top off a similarly sized regular egg, tip out its contents then gently place the cloned embryo inside with a little egg white before sealing it up again. No one as yet has tried this last stage with a cloned avian embryo, but Kjelland has sliced the top off a turkey egg and had a regular Rhode Island Red chick hatch from it; proof in principle that the embryo transfer technique will work.

The component parts of the avian cloning process, it seems, are pretty much there, but they need to be refined and practised by skilled, technical hands, then put together. No one has, as yet, done all of the steps from start to finish. 'The main problem now,' says Kjelland, 'seems to be a lack of funding and a defeatist attitude. We need to overcome that.' But in the meantime, there may be another method, more developed and well rehearsed, that could be used to make a baby passenger pigeon …

Don't Cry, Daddy

Poor old dads. When a baby is born, its father seldom receives much in the way of recognition. All eyes are on the little one, wrinkled and vulnerable, and on the mother, stoic and resilient, justly exhausted after the Herculean tasks of being pregnant and giving birth. Poor, neglected, unnoticed fathers. No one acknowledges how well they get on with their daily lives as their partner's belly balloons and her ankles swell. No one congratulates them for eating and peeing normally, while their partner wretches over the sight of toast and struggles with a bladder the size of a pea. Few pat their backs as they stand around the labour suite immersing themselves selflessly in the role of 'spare part' while their partner painfully attempts to push a 'square peg' through a 'round hole'. Poor old dads.

New life is created every minute of every day, yet dads rack up little in the way of credit. Then, every once in a while, comes a story that celebrates the role played by the father, a tale that embraces his very unique role in the miracle of creation. This is one of those stories.

In March 2013, a very feathery father bagged news column inches around the world. This was one dad who really had done something quite extraordinary. 'Duck fathers chicken' shouted one headline. 'Chicken has a duck for a father' yelled another. 'Chick dad is quackers' mumbled another apologetically under its breath.* If ever there was a case for a paternity test, this was it. In some bizarre, laboratory-fuelled feathery tryst, a male duck and a female chicken somehow got it together and managed to produce, not a dicken or a chuck, but a pure-bred baby chick. The new arrival was announced in the peer-reviewed journal *Biology of Reproduction*, which published a heart-warming photograph of mother hen and daddy drake looking mighty pleased with themselves as they flanked their fluffy, yellow chick against a sunshine-dappled lawn. But this was no ordinary dad. This was a duck with a secret. A secret so secret, not even he knew about. A secret that he kept ... in his gonads.

A year or so earlier, when the duck, known un-affectionately as 'wd25', was but a tiny embryo he underwent a procedure in a Dubai laboratory. Cell biologist Il-Kuk Chang and colleagues from the Central Veterinary Research Laboratory injected chicken cells into his bloodstream. But these were no ordinary cells. They were primordial germ cells or PGCs, specialised cells formed early in development that do just one thing – sit in the gonads and make sex cells. In the female they make eggs, and in the male they make sperm.

*It didn't but it should have done. The sub-editors missed a trick.

When Chang injected the chicken PGCs into wd25's aorta, they travelled through his blood supply to his gonads, where they bedded down among the duck's own PGCs and started to make sperm – chicken sperm. With a mix of duck and chicken PGCs, when the drake reached sexual maturity, he started to make both duck and chicken sperm. According to the research paper, 'semen samples were collected* and then used to inseminate female chickens, who then went on to do what chickens do best: lay lots of eggs. Of these, a handful hatched into *bona fide* chicken chicks – healthy, baby birds that were genetically all chicken, but with a duck father and a chicken mother. Talk about giving a youngster an identity crisis.

It's an eye-catching piece of science, but one with a much deeper purpose. If the eggs and sperm of one species can be grown inside the reproductive system of another, then the technique could be used to help boost the numbers of endangered species and potentially bring back extinct ones. PGCs from band-tailed pigeons, edited to contain passenger pigeon DNA, could be injected into a host bird embryo. This, most likely, would be a band-tailed pigeon, which would then be allowed to grow up as normal and which would, in time, begin producing passenger pigeon sperm. If female band-tailed pigeons could be similarly altered to produce passenger pigeon eggs, then in theory at least, passenger pigeons could be bred.

Surrogate parents, most likely band-tailed pigeons, could be used to incubate the eggs and look after the hatchlings. But given that band-tailed pigeons are almost monochrome, and that passenger pigeons were multi-coloured, Ben Novak tells me that the surrogates might well have their feathers dyed to make them more convincing. 'After all,' he says, 'there's no such thing as getting it too right.' But a baby

*The mind boggles … I like to think there was mood music and a copy of 'Mallards' Wives' involved.

passenger pigeon might not be expecting much in the way of parental care. Historical records reveal how all the members of a single flock would nest in the same forest at the same time, constructing millions of rickety nests high up in the tree tops. After that, they'd all lay their eggs – one or two per nest, pretty much in unison – then incubate them for a couple of weeks. When the squabs hatched, the parents would feed them 'pigeon milk', a fatty, protein-rich paste they produced in their crops, along with bits of regurgitated food. Then, at the tender age of just two weeks old, something incredible happened. The parents gave their babies one last feed, then rose up together and disappeared. They just flew the coop. The helpless baby birds were left all alone, three metres (10ft) up, without food and with stumpy feathers that had yet to morph into wings. Bemused, the increasingly ravenous babies would stay in the nest for a day or two then, one by one, flop down to the forest floor in search of food. It's like tipping a toddler out of a pram and hoping it can blend its own baby mush. In the few days it would then take for their wings to develop, the grounded youngsters were easy pickings for woodland carnivores. Wolves, foxes, hawks and other predators had a field day. It's like tipping a toddler out of a pram and hoping it can blend its own baby mush … in a kitchen full of baby-eating tigers. Someone call Avian Social Services.

It seems like a cruel and wasteful strategy, but the likelihood is, with their food reserves dwindling, the adults simply had to move on in search of sustenance. Sure, their desertion led to the deaths of many thousands of chicks, but in evolutionary terms the damage was collateral. The colony as a whole survived to see another day. And odd though it may seem, this lacklustre style of parenting could work in our favour today. Because infant passenger pigeons were simply used to 'getting on with it', a newly de-extincted bird may be unlikely to ask much of their surrogate parents beyond a little crop-milk and some regurgitated seeds.

The plan would be to raise the birds carefully in captivity, in as large an aviary as possible, until they reached flock-sized numbers and could be released into the wild. The deciduous forests of eastern North America, once felled by European farmers, have largely regrown, so there's habitat to be had – although, some argue, maybe not enough. As nomadic birds that followed their beaks to find food rather than retracing defined migratory routes, they shouldn't need to be taught where to go. Instead, if they're anything like their forebears, they'll instinctively flock together and forage, searching out nut-rich woodland and perhaps even, seed-rich farmland. 'The juveniles will likely form their own social structure,' says Novak. In terms of numbers, Revive and Restore are aiming for an as yet unknown quantity, but one likely to be in the hundreds of thousands, perhaps the millions. It might sound like a lot, but many birds exist in these kinds of numbers and above. In 2014, there were an estimated 275 million mourning doves. So there's plenty of room for the passenger pigeon.

The key for Novak, and indeed for Revive and Restore, is to recreate a bird that fulfils the same ecological niche as its forebears. In the passenger pigeon's case the aim is to make a twenty-first-century bird that behaves like a passenger pigeon and interacts with its environment in the same kind of way. If Novak does his job right, the revived and restored passenger pigeon will roam in dense flocks,* eating nuts and seeds, destroying some, dispersing others and helping to enhance woodland diversity. It will be a source of protein for carnivores and a source of competition for other fruit and grain-eaters. However, all this will take time. With a lot of basic science still to do, Revive and Restore doesn't expect to hear the tiny pitter-patter of passenger pigeon

*I've presumed, throughout this chapter, that a 'flock' would be the correct collective noun for passenger pigeons, but perhaps they deserve their own term. I suggest either a 'tempest', an 'eclipse' or a 'Novak' of passenger pigeons.

feet before 2022, and envisage the first test flocks could be flying by 2032. So, not imminent, but keep watching the skies.

The project is not without its sceptics. Some, such as Beth Shapiro, cite technical hurdles. 'We can't de-extinct the passenger pigeon at the moment,' she says, 'for lots of reasons.' No one, as yet, has made or modified pigeon PGCs, much less transferred them into a host bird to have it make sperm or egg, and then used those sex cells to generate live birds. But I've worked as a cell biologist. I've tinkered with genetic modification. I've spent long, frustrating days trying to fathom the exact nutrients needed to keep my petri dishes of cultured cells alive. I've curled my fists in frustration as my primitive attempts to genetically edit said cells yielded unspectacular, unsatisfactory results. I know how difficult it can be to work with cells and with DNA, but with the benefit of 20 years' hindsight, I can see just how far and fast the technology is progressing. Yes, there are hurdles, but with enthusiastic, clever people like Ben Novak at the coal face, there's no reason these hurdles won't be overcome.

But just because we can do something, doesn't automatically mean we should. Ornithologist Mark Avery, author of *A Message from Martha*, is concerned about the birds' welfare should we bring them back. In the early days of their de-extinction there would, unavoidably, be small numbers of birds raised and bred in captivity. 'But we know that these were highly social birds,' he says, 'so it all feels rather sad.' These were birds that flocked in the millions. They needed to. It offered them safety in numbers and buffered the population when so many of their unattended young fell to Earth and got eaten. We can't bring back and release a handful of passenger pigeons. They wouldn't stand a chance. With the passenger pigeon it's all or nothing. We either bring back a flock so large it would darken the sky, or we don't bother at all. And it's here that I find myself erring on the side of caution.

In 1855, one witness to a passenger pigeon flock described how 'children screamed and ran for home. Women gathered their long skirts and hurried for the shelter of stores. Horses bolted. A few people mumbled frightened words about the approach of the millennium, and several dropped on their knees and prayed.' These were rowdy, riotous birds that left devastation in their wake. They could devour the contents of a newly planted field in minutes. They ruined entire harvests and pissed off farmers on a monumental scale. When they flew down to roost, they'd perch on any and every available branch. If there wasn't any space, they'd pile on top of one another, causing boughs to break, crushing any birds on the branches below. When they deposited their droppings it was like giant snowflakes falling from the sky, leaving the ground covered in a layer of deep guano. And when they moved on, they left behind a landscape so apocalyptic it was worse even than the aftermath of a toddler's birthday party. The great naturalist James Audubon said it was 'as if the forest had been swept by a tornado.' Opportunistic omnivores, they flew from place to place like feathered locusts devouring whatever they could. And if they ate one thing then found another they liked better, they'd throw up the contents of their crops and keep on bingeing. Passenger pigeons were the hell-raising, feathery bulimics of North America.

To choose as a de-extinction candidate a species whose survival very probably depends on it existing in such vast numbers is, I think, a controversial choice. Perhaps we'd end up having to cull the creatures we went to so much trouble to create. 'It could be like Martha all over again,' says Avery. Given that pigeons aren't exactly the most popular birds alive on the planet today, how ready are we, I wonder, for the passenger pigeon?

CHAPTER FIVE

King of Down Under

It was bitterly cold that night. As the temperature slid silently towards zero, a caged thylacine paced up and down under the starry Tasmanian sky. Numbed by the cold, sinking slowly into hypothermia, the stripy, wolf-like beast cried out. But there was no one around to listen, no one to care. The keepers had long since locked up and gone home.

It was 7 September 1936, and the height of the Great Depression. At the Beaumaris Zoo in Hobart, times were hard. To keep costs down the zoo employed disinterested cheap labour – 'sussos', as they were known – to look after the animals. They worked for minimal wages with minimal supervision. As a result, cages were left uncleaned and animals were left to pick over the rotting remains of previous meals. Their most basic needs neglected, many of them became sick, but no one bothered to call the vet. Vets cost money.

The thylacine, who was known as Benjamin, lived in a sparse, rectangular enclosure at the back of the zoo. Although it was the start of spring, the daytime had been unseasonably hot. The single tree that covered Benjamin's enclosure had yet to sprout leaves, and the thylacine had been left without shade. Visitors came. Visitors went, but Benjamin was too dehydrated to take much notice. At dusk, the keepers were meant to open a sliding wooden door and shoo the animal into its covered nighttime pen, but they forgot … again. The exhausted thylacine was left alone to face the elements. While his keepers slept in their beds, Benjamin was unable to get to his. And when they turned up for work the next day, Benjamin was dead. In a final act of indifference, they threw his cold, lifeless body out with the trash.

It was a sad, pointless death that could so easily have been avoided, but made all the more poignant by the fact that Benjamin was irreplaceable. He wasn't just the last thylacine in the zoo. He was the last known thylacine on Earth.

Thylacines, known also as Tasmanian tigers and Tasmanian wolves, were enigmatic and unusual animals. Think of a large dog with pointy ears, wearing black eyeliner and a tiger onesie. It had the head of a wolf, the stripes of a tiger and the stiff tail and pouch of a kangaroo. People said it walked like a 'dog with a broken back', yet it could turn on the speed and disappear into the bush in the blink of an eye. A fierce predator, it fed on wombats, rodents and birds. Notoriously vocal, it made a variety of bizarre noises: hisses, snuffles and a particular wheezy rattle that made it sound like it needed an inhaler. It looked like a placental mammal, yet was a marsupial. But where most marsupials have pouches that open upwards, the thylacine had one that pointed down.[*] Unusually, both sexes had a pouch. Females used theirs to nurture their babies, while males used theirs to protect their pendulous, dangling scrotums from the prickly Tasmanian scrub[†].

They first appeared around 25 million years ago. Thylacine fossils have been found at the Riversleigh World Heritage Site in Queensland, Australia, alongside the remains of other fantastical beasts: carnivorous kangaroos; tree-climbing crocodiles; and one of the largest birds ever found, a flesh-eater dubbed the 'demon duck of doom' (*Bullockornis planei*). Back then there were at least six different types of thylacine, some as big as Labradors, others as small as Chihuahuas. 'Paris Hilton would have been able to carry one of these things around in a little handbag,' palaeontologist Michael Archer from Australia's University of New South Wales told the 2013 TEDx De-extinction event, 'until a drop-croc landed on her.' Then gradually, the world began to change. It became cooler and drier, and the different types of thylacines struggled to adapt. By four million years ago, they had been whittled down to just one, the modern thylacine, *Thylacinus cynocephalus*, 'the pouched dog with a wolf-like head'. Then it, too, began to disappear.

Sixty thousand years ago, Australia, New Guinea and Tasmania were part of a single landmass, still home to the

---

[*]Water opossums (*Chironectes minimus*) also have downward-facing pouches. Alive and well today in the freshwater streams and lakes of Central and South America, the water opossum's pouch can be closed and made watertight by a strong ring of muscle. Babies get to stay dry, while the males' genitalia, hidden therein, can't get tangled up in underwater plants. Phew!

[†]The first scientific description of the thylacine, recorded in 1805 by Tasmania's Lieutenant-Governor Paterson, described the amply endowed thylacine thus: 'Scrotum pendulous, but partly concealed in a small cavity or pouch in the abdomen … Eyes large and black … which gives the animal a savage and malicious appearance.'

thylacine. But then humans arrived from Asia. Where the thylacine had been an apex predator for millions of years it now found it had competition. By hunting the same animals as the thylacine, humans began to nudge the delicate balance of the Australasian ecosystem out of kilter. The landmass split. By 10,000 years ago, thylacines were extinct in New Guinea. Then, 3,500 years ago, new colonists arrived in Australia and brought dingoes with them. Used as a hunting companion, the pack animals helped humans become even better at killing things. Out-competed and out-manoeuvred, thylacine numbers declined until, one day, the Tasmanian tiger was gone from mainland Australia, too.

Tasmania was the thylacine's last stronghold, but even that was to fall. In the early nineteenth century, settlers arrived from Europe, determined to turn Tasmania into a little slice of home. They built churches, houses and sheep farms, and replaced native animals with European alternatives. Rumours began to circulate that the thylacine was a sheep killer, and from a small Chinese whisper a larger wave of paranoia swept the island. If the gossip was to be believed, thylacines not only killed sheep, they sometimes took small children too. In the end, people were so frightened that the Tasmanian government placed a bounty on the thylacine's head. Every man with a gun turned against them. Thylacines were slaughtered in their thousands.

The tragedy is that the rumours were unfounded. There's no concrete evidence to suggest that thylacines ever took children. The stories were folklore, most likely made up to ensure that curious kids kept their distance. Nor is there any evidence to suggest that thylacines killed more than just the odd sheep. They certainly weren't the blood-thirsty murderers they were made out to be. The truth, according to biologist Robert Paddle from the Australian Catholic University (author of *The Last Tasmanian Tiger: The History and Extinction of the Thylacine*), is that thylacines were used as scapegoats. Sheep farms failed, not because thylacines killed sheep, but because of bad weather and poor management. Rather than admit their failings, it was easier for the middlemen of Tasmania's sheep industry to blame it all on someone, or something, else. Thylacines became the fall guy and then paid the price.

By the early 1900s, thylacines were incredibly rare in Tasmania, and by 1936, there was just one left – Benjamin. Three years before he died, Benjamin was filmed in his outdoor enclosure by Australian naturalist David Fleay. It's one of the most haunting pieces of footage you'll ever see. Google it, I urge you. The film is washed out and grainy. There's no sound. A black and white Benjamin paces around his small, sparse enclosure. He eyes the camera and seems to yawn. He looks like he's given up already. The film loops and repeats. But it's what happened after the camera stopped rolling that, for me, brings Benjamin to life. Thylacines, you see, 'yawn' when they feel threatened. That staggering 120-degree gape was a sign that they were not happy, that if pushed, they were likely to attack. A black box on a tripod with a curtain and a pair of legs sticking out was definitely threatening, and Fleay either ignored or was unaware of the warning sign. When the camera stopped rolling, the last living thylacine on planet Earth bit the cameraman on the arse – not once, but twice! Thylacine 2 – *Homo sapiens* 0! It's a wonderfully defiant glint of life at the conclusion of an otherwise tragic and sorry story.

Ironically, the thylacine finally received full legal protection in the year that Benjamin died, but it was much too little, much too late. Since then it has become a much-missed Tasmanian icon and many people refuse to believe it has gone. Since Benjamin's death, there have been over 4,000 reported sightings of possible thylacines. Roberta Westbrook, the landlady of the Mole Creek Hotel in Northern Tasmania, claims she saw one in 1997 when she was driving along the road from Mole Creek to Paradise.

Its eyes, she says, were dark, like the animal wore eyeliner. Then in 2010 a French backpacker spotted a similar animal on the same stretch of road. A kohl-pencilled fox or genuine thylacine? On the remote western side of New Guinea, locals talk of a thylacine-like creature they call the *dobsegna*, which has been spotted as recently as 1997. And from Australia, there are grainy photos and blurred videos. One piece of film, shot from a car in 1973, shows an apparently striped, dog-sized animal run out of some trees and cross a road. Its gait is part dog, part kangaroo and its tail is held out stiffly. But in the blink of an eye, the animal is gone.

The problem, however, is that the 'evidence', if one can call it that, is either equivocal or downright damning. Some photos and films are proven fakes, others too fuzzy to make an unambiguous ID. Stories are just that, while eyewitness testimonies come without hard proof. Samples of suspected thylacine hair and droppings have all turned out to come from other native wildlife. One study by the Queensland Museum's Ralph Molnar found that patterns of thylacine sightings fail to match those of other Australian wildlife, as would be expected, but do mirror sightings of UFOs! Just like flying saucers, sightings of thylacines tend to be made by single individuals, last just a few minutes and occur late at night, often after the pub has shut. These days we have camera phones glued to our fingertips, yet nobody has been able to take an image good enough to convince the doubters. In the absence of a body or DNA proof, the thylacine, in my eyes at least, remains resolutely and sadly dead. Why chase rainbows when something practical can be done?

In a Pickle

An Australian by birth, Michael Archer grew up in the Appalachian Mountains in the eastern United States. As a child, he was more interested in the local wildlife than he was his fellow classmates. 'I was a bit of a sociopath,' he says. 'I didn't enjoy the company of people that much; snakes and turtles seemed much more reasonable to me.' Then he discovered fossils in the rocks and boulders around his home, and a lifelong interest in the lives of animals long gone was born. He collected and stored all that he found in a special room in his house. Then one day, on a visit to New York, he heaved two fossil-filled suitcases up to the front desk of the American Museum of Natural History and asked someone to take a look. The late curator of invertebrates, Norman D. Newell, graciously obliged and identified his specimens for him. It was an act of kindness that inspired Archer to follow a career in science. He studied Geology and Biology at Princeton then, while a Fulbright Scholar at the Western Australian Museum in Perth, became intrigued by the museum's collection of pickled Australian animals. Many of the specimens, he realised, had never been properly identified. 'They were unknown to science,' he says. It made Archer realise just how little was known about this cache of Australian biodiversity. For his PhD at the University of Western Australia, he studied carnivorous marsupials, including the thylacine, and went on, through much of his career, to catalogue many of the amazing finds from Riversleigh. But it was a visit to the Australian Museum in Sydney in 1976 that would change everything.

'Staring down at me from a shelf full of skulls and skeletal tissue was a thylacine pup in a glass jar,' he says. The pup had been acquired over a hundred years before by museum curator and collector George Masters. A Victorian 'Crocodile Dundee', Masters travelled widely through Australia and Tasmania, wrestling venomous snakes and shooting all manner of fauna to bring back for the collections at the Australian Museum. At one stage he had collected more than half of the museum's natural history exhibits, and the thylacine was among them. Exposure to daylight and the alcohol it had been preserved in had long since bleached most of the colour from the tiny thylacine's body, so the pup was white as a sheet. Only the faintest of stripes could be seen on its back. It lay curled up with its tail tucked under its bottom, its forepaws curled up as if begging. Wrinkly folds of skin hung loose over its podgy belly and its eyes were squeezed shut. Archer was mesmerised.

In the years that followed, he went to see the pup several more times. Because of his research interests, Archer realised better than most the pivotal role played by humans in the thylacine's extinction. 'We killed these things,' he tells me. 'We slaughtered them. What I think is important is that, if it's clear that we exterminated these species, then I think we not only have a moral obligation to see what we can do about it, but I think we've got a moral imperative to try to do something if we can.' The ghostly thylacine pup haunted him. Then, during one visit in 1990, a thought popped into his head. Archer knew enough about preservation to realise that alcohol preserves not just cells, but the DNA inside them too. If the pup had been pickled promptly after its death, perhaps its DNA was still in good shape and could be used for cloning. Perhaps mankind could undo some of the wrongs it had done and de-extinct the thylacine.

It was an audacious idea. In 1990, claims that DNA had been retrieved from ancient tissue samples were largely met with scepticism, and mammals had yet to be cloned. Archer asked around his geneticist friends to see what they thought. 'There was uproarious laughter,' says Archer. 'Colleagues thought the idea was completely ridiculous.' But he wasn't put off. He decided that if he ever got the chance, he would give it a go.

The opportunity came around a decade later when Archer was appointed Director of the Australian Museum, home of the pickled thylacine. On his orders, the pickled pup was removed from its ethanol bath and a tiny piece of tissue removed. Geneticists Don Colgan and Karen

Firestone then ran tests to see if the tissue still contained DNA. Much to their delight, they found that it did.

The plan was to create a 'living library' of thylacine DNA, where recovered genetic material would be stored in living bacteria that could be kept in the lab. From this Archer hoped to amass material for cloning. The thylacine genome would be reconstructed inside a cell belonging to the animal's closest living relative, the Tasmanian devil (*Sarcophilus harrisii*), and if a cloned embryo could be made, it would be nurtured using the surrogacy skills of the same species. Beyond that, Archer envisaged a scenario where cloned thylacines would be raised in captivity, allowed to breed naturally, then released back into the wild. 'There's plenty of places in Tasmania where they could still live,' he says. On one visit, Archer went to the beech forests on the island's south side where a local man, Peter Carter, who remembered the animals from his childhood, showed Archer around. Carter told him how the thylacines used to circle his old hunting shack at night, and how when he was a boy, he was allowed to keep one on a lead. Back in the day, despite all the erroneous rumours of sheep killing and child snatching, thylacines, it seems, were kept as pets. Some museum specimens bear rough rings of fur round the neck where their collars rubbed. And historical records reveal how in 1831, Tasmania's first ever shop, a livery stable in Hobart, had a live pet thylacine for sale. Archer sees no reason why de-extincted thylacines couldn't be kept as pets today. 'We're at a situation where, increasingly, wildlife isn't safe in the wild anymore,' he says. 'We need additional strategies to help protect what we've got.' It's a controversial idea. Many people won't be happy with the idea of de-extincting a wild animal only to have it live as a lapdog, but Archer's idea would be to have pet thylacines *and* captive and wild populations. It's a natural leap for him to make. Over the years, he's kept all manner of Australian wildlife as pets in his home, including wallabies, quolls, possums and fruit bats. 'Marsupials,' he says passionately, 'make excellent pets.' But the purpose wouldn't be companionship. It would be to raise the thylacine's profile, to make people care more about wildlife and the rate of its disappearance. 'No animal we've ever put our arm around has ever gone extinct,' he says. But one step at a time …

Shortly after his team extracted snippets of DNA from the thylacine pup, Archer fronted a press conference where he told the world's media 'Ladies and gentlemen, we are here to announce what is probably the biological equivalent of human beings taking their first step on the moon.' It was a bold, headline-grabbing statement, and one that I think, with the benefit of hindsight, Archer has come to regret. From that moment on, the world's media was captivated. Journalists badgered Archer and his team for constant updates and the Discovery channel filmed their every move. It created a lot of pressure, but the media didn't grasp the subtleties of what was going on. Where before the team had isolated bitty, fragmented strands of DNA, they now found themselves able to retrieve whole genes, not just from the pickled pup but from other dried-up specimens. It was a significant step but all the media saw was a big, fat absence of a living, breathing thylacine. 'That was the biggest problem with the thylacine project,' says Archer. 'From day one, it was conducted in the full glare of the media spotlight.' Then, in 2003, Archer left the Australian Museum to become Dean of Science at the University of New South Wales, where he can still be found today. At the Australian Museum, the thylacine project fell by the wayside. 'I never expected that the thylacine project wouldn't continue without me,' says Archer. 'It was a huge disappointment.'

Since then, no one has picked up where Archer left off. There is, at present, no coherent plan to de-extinct the thylacine. But relevant research is continuing to build. The thylacine's mitochondrial genome has been published, and there's evidence to suggest that DNA sequences from preserved specimens might still actually work. In 2008,

Andrew Pask from the University of Melbourne and colleagues extracted a tiny snippet of DNA from some century-old pickled thylacines. The fragment wasn't a gene but a stretch of DNA that switches on the gene that codes for collagen, the structural protein found in bone and cartilage. They joined the fragment to another gene that produces a blue pigment, and injected the hybrid DNA into developing mouse embryos. Fourteen days later, when the embryonic mice were processed, they were very blue indeed, hinting that the thylacine DNA had managed to switch on the rodent collagen gene. Pask may not have resurrected the thylacine, but he did bring its DNA back to life. That has to be a step in the right direction. For now, it's as close to a living thylacine as anyone has got, but Archer is not disheartened. He has another de-extinction project on the go.

A Frog in the Throat

The Lazarus Project seeks to bring back one of the most bizarre animals in the world. It's everything the thylacine isn't. This creature wasn't furry or pouched, big or stripy. It didn't have sharp pointy teeth or a long tail. It never inspired fear or fable. The truth is, few people outside of its native homeland have ever heard of it. It never bothered sheep, people or anything very much, but it might have hurt a fly. The animal was a humble, slimy frog that lived, until very recently, in the bubbling creeks of Queensland, Australia. But it had one hell of a party trick. Wait for it … drum roll please … female frogs burped up fully formed froglets. I kid you not, they really did.

Think for a moment how peculiar that is. Female frogs, we are told at school, ditch their blobby frogspawn in a ditch or pond. After the eggs have been fertilised by a male, they hatch into tadpoles, which then turn into frogs. It's a process that plays out in jam jars and buckets all over the world, much to the wonder of transfixed children.

That's how frogs have their young. Mother frogs do not, instinct tells us, belch up froglets. Burping releases gas, not babies. If it were any other way I'd pop out quadruplets every time I had a gin and tonic. Thankfully this is not the case, yet here is an animal that somehow ejects fully formed froglets from its mouth. It gives birth by burping.

Now, I've experienced giving birth via the conventional route … several times. On one occasion I even had to push out two whoppers in the space of an hour. But it's not pretty. In that short time, I experienced pain that was off the Richter scale, communicated only in words of four letters, and nearly punched the midwife. Repeatedly. The babies came out beautifully but my vestibule* was left in tatters and any dignity I'd once had was washed away when the tsunami that was my waters broke. Burping up a baby, in contrast, seems like an excellent idea. Who hasn't, at some point, enjoyed the satisfaction of a loud, resonating belch? Be honest. The freedom! The release! The production of a sound so at odds with your normal voice it shouldn't be possible. But it is. A frog that belches and gives birth at the same time is nothing short of an evolutionary masterstroke.

The southern gastric-brooding frog (*Rheobatrachus silus*) was discovered in 1973, the same year that Elvis Presley's 'Aloha from Hawaii Via Satellite' concert attracted more viewers than the Apollo moon landings. It was chanced upon by biologist David Liem during a routine field survey of the rocky streams of south-east Queensland. As frogs go, it wasn't much of a looker. Unremarkable. Drab. Not particularly big. Not particularly small. But Liem realised it was different to other Aussie frogs. It had large, bulging eyes, a short, blunt snout and an excessively slimy body that made the animal difficult to handle. To Liem it looked more like a frog found in Africa, *Xenopus laevis*, than it did

*See Chapter 3 for an explanation of terminology.

any native amphibian. Yet here it was, hiding among the stones of a fast-flowing Australian creek. Unaware of its unusual talent, Liem's colleagues, biologists Chris Corpen and Greg Roberts, captured a few of the animals and because it was getting late, decided to take them back to their house for an overnight stay. They would take them to the lab later.

As house guests go, they made quite an impression. The amphibians were in a tank in the living room when one of the housemates noticed that a big frog seemed to be eating a smaller one. But when the researchers looked more closely, they realised that the little amphibian was coming out of, not going into, the bigger frog's mouth. It was all a bit odd. The scientists were impressed but not totally flabbergasted. After all, male Darwin's frogs (*Rhinoderma darwinii*) are known to rear and transport their tadpoles in enlarged vocal sacs. Perhaps, the researchers reasoned, this was a male frog that was doing something similar.

Then it happened again. The frogs had been transported to the lab and one of them was to be moved to a separate tank. But when the researcher plunged his hand into the water, the frog slipped through his fingers, rose to the surface and much to everyone's surprise, burped up not one, but six little froglets. Over the next few weeks, three more tiny frogs appeared, so the researchers decided to examine the parent frog more closely. But when they tried to pick it up, the frog raised its head and belched up octuplets. Then, without warning or time to call the midwife, five more froglets followed. When team leader Mike Tyler at the University of Adelaide and colleagues were finally able to dissect the frog, they found the animal they had presumed was male was in fact female. The mother hadn't burped up her offspring from a vocal sac, but instead had nurtured them inside her very big, very stretchy stomach.

The gastric-brooding frog was quite unique. A female frog, we now know, would lay her eggs as normal, then swallow them once they had been fertilised. The eggs would then slip down into her stomach, where they would hatch into tadpoles and spend the next six weeks metamorphosing into frogs. A pregnant mother could fit a staggering 20 froglets inside her TARDIS-like body, an impressive feat given that she was just 7cm (2.8in) long, and each full-term froglet measured 1cm (0.8in) in length. As her babies grew, her stomach became stretched as thin as a plastic sandwich bag, filling her body cavity until her lungs became so squished she had to breathe through her skin. With her buoyancy and centre of gravity altered, the bloated female could no longer float horizontally and instead was forced to dangle vertically in the water with her head sticking out. Then when she was ready, she'd simply burp up her babies, one or more at a time, as and when they were ready. And after that, her stomach would return to normal as if nothing out of the ordinary had ever happened.

Mike Tyler and colleagues described this 'unique form of parental care' in the journal *Science* in 1974, but it didn't receive the fanfare that was expected. Instead of awe and wonder, the paper was met with scepticism and disbelief. Gastric-brooding, it seemed, was an outlandish, impossible idea. A stomach can't turn into a womb any more than a womb can turn into a stomach. Tyler, critics proclaimed, must have got it all wrong.

So Tyler set about studying and photographing the frog further until he had amassed enough data for a second, more detailed research paper that described both the animal and its behaviour. This time he included his photographic evidence. The most iconic image is of a female held snuggly between her captor's thumb and forefinger, her mouth wide open with a mini frog climbing out. It looks like the ugliest Russian doll you've ever seen, but it was irrefutable. When Tyler published the second paper in 1981, the response was completely different. The baby-burping gastric-brooding frog was featured in newspapers, magazines and journals around the world. Several groups began to study the frog.

But just as interest in the gastric-brooding frog was mounting, its numbers in the wild were waning dramatically. Tyler and his team visited the creeks of south-east Queensland every month between 1976 and 1980, but the little frog became harder and harder to find. The last wild gastric-brooding frog was spotted in 1981, and although people continued to look for it, it has never been found since. After that, the last two adult frogs that Tyler had kept in his laboratory died in 1983 and the species became officially extinct. Just 10 years after it had first been spotted, the southern gastric-brooding frog was no more. What a loss.

Then, on New Year's Eve 1984, some good news! A man who would later become known as 'The Frog Whisperer' for his ability to find and catch wild frogs, was taking a dip in a waterfall, high up in the mountains of Queensland's Eungella National Park. Biologist Michael Mahony, then at Sydney's Macquarie University, was taking a well-earned break after a hot, humid day spent collecting frogs for his PhD project when one of his colleagues spotted an orangey-brown frog disappear under a stone. From his knowledge of the local fauna, Mahony knew this had to be something unusual – the only orangey-brown amphibian around those parts was the cane toad, but that wasn't found in mountain streams. He felt around under the rocks and clasped his hands around it, but the creature was unusually slimy, a trademark of the slippery gastric-brooding frog. 'So I knew what it was before I saw it,' he says. Mahony had discovered a second species of gastric-brooding frog, the northern variety (*Rheobatrachus vitellinus*).

These frogs were not difficult to find. Mahony and colleagues were able to collect dozens of them in just one night. Hopes were high that the animal could be studied and the secrets of its gastric-brooding unravelled. 'Finding the new species of gastric-brooding frog was like a second chance,' says Mahony, 'but it was bittersweet.' Almost as soon as it was discovered, the northern gastric-brooding frog disappeared too. Queensland scientists, charged with monitoring the species, watched as local populations crashed. 'There were no proactive efforts to save the frog,' says Mahony, 'the scientists effectively monitored it into extinction.' Less than a year after Mahony identified the slippery first specimen, the northern gastric-brooding frog disappeared from the streams of Eungella. It too had become extinct.

The truth of it is that we humans have inadvertently been spreading mass 'frogicide' around the globe in the shape of a nasty fungus that infects and kills frogs and other amphibians. Disseminated around the world via the commercial trade in amphibians for the pet and food industries, the chytrid fungus (*Batrachochytrium dendrobatidis*) is now found on every continent where there are amphibians (that's everywhere except Antarctica). It enters the animals' bodies through their skin, upsets their fluid balance and kills by causing heart failure. Chytrid has caused mass mortalities, population declines and extinctions on multiple continents, and in terms of hammering biodiversity is the most significant disease of vertebrates in recorded history. Over the past 30 years it has caused the catastrophic decline or extinction of at least 200 species of frogs. Although the fungus wasn't formerly recognised until 1999, analysis of a preserved museum specimen suggests that amphibians living in the gastric-brooding frog's backyard – the mountain ranges of south-east Queensland – were infected with the fungus as early as 1978. The last wild southern gastric-brooding frog was spotted just three years later. 'We think gastric-brooding frogs were highly susceptible to chytrid,' says Mahony. Once again, it seems we humans are responsible for the disappearance of yet another species.

But Michael Archer hopes to change all that. Under the auspices of the Lazarus Project, he has amassed a formidable team of experts, including Michael Mahony and Mike Tyler, to de-extinct the gastric-brooding frog. As with the thylacine project, Archer sees the Lazarus Project as an opportunity for mankind to atone for its crimes against biodiversity, but it's also more than that. Like the thylacine, the gastric-brooding frog represents something quite unique in evolutionary terms. There's an argument to be had for bringing it back purely because it is so unusual. It's physically and genetically distinct. The female's ability to turn its stomach into a makeshift womb sets it apart from anything else that is alive on the planet today. If we could de-extinct the gastric-brooding frog, then we'd have an opportunity to understand how these changes occur, to explain how it is that female frogs don't digest the eggs that they swallow. Preliminary studies from when the frogs were still alive suggest that the eggs' jelly coats contained a substance, prostaglandin E2, which dampened the production of stomach acid, and that when the tadpoles hatched they produced the same hormone-like substance. Archer argues that if we de-extincted the gastric-brooding frog and worked out how it managed its gastric acid secretions, then it may lead to the development of new therapies for stomach ulcers or for people recovering from stomach surgery. But first Archer has to clone it.

Early Days

Luckily for the Lazarus Project, other types of frogs have already been cloned. Back in the fifties, people weren't interested in cloning frogs *per se*, but in a question that had been nagging cell biologists for over half a century: what happens to the genome of a fertilised egg as it starts to develop and turns, ultimately, into an adult animal? In the very early days of embryonic life, the cells inside an embryo are totipotent, meaning they can turn into any of the many hundreds of different cell types found in the adult animal – heart, muscle and nerve cells, for instance. But as the embryo develops from a seemingly unimpressive blob of cells into something more recognisable, with limbs, organs, tissues and the like, the cells start to lose their totipotency. They become less able to generate different cell types. So the question that kept cell biologists up at night was, as an embryo develops to adulthood, do its genes stay the same but are somehow switched on and off differently, or are the parts of the genome that are no longer needed somehow lost? Nuclear transfer experiments held the answer. If the DNA-containing nucleus of an adult cell could be transferred into an egg stripped of its own genetic material, what would happen? If development proceeded as normal, then the requisite genes must have been reactivated. But if the cloned egg failed to divide, perhaps the DNA needed for development had somehow been lost forever …

Scientists had been trying and failing to take a nucleus from one cell and put it into another, physically separate cell, but the cells broke so easily that most people thought the task was simply impossible. So when one man, Robert Briggs, applied for funding from the National Cancer Institute (NCI) for his nuclear transfer experiments, the 'hare-brained scheme' was roundly dismissed. But Briggs, a banjo-playing ex-shoe factory worker, didn't give up. The NCI relented and Briggs acquired the assistance of a talented researcher called Thomas King. For their work, they chose to focus on a beauty of a frog called *Rana pipiens*, the northern leopard frog. With its green body and striking black spots, it looks like it has hopped straight out of a colouring book. Briggs and King took a single cell from an embryonic frog, sucked out its nucleus with a fine glass pipette, then injected it into an egg that had had its own nucleus removed. It was frustratingly slow work and the duo spent two years honing their methods and gathering data. But what they found was that 40 per cent of the DNA-infused eggs developed into embryos and then tadpoles. It was the first *de facto* nuclear transfer experiment, and a huge technical breakthrough. They published their research in 1952 to much scientific acclaim. But while it sort of answered the question they were interested in, it didn't quite nail it. Briggs and King had shown that DNA from embryonic frog cells could direct development, but what of DNA from older, more specialised cells? It would take a man who would later become known as the 'Godfather of Cloning' to sort that one out.

At school, British-born John Gurdon was advised against becoming a scientist. In his end-of-term report his teacher wrote, 'I believe Gurdon has ideas about becoming a scientist. On present showing, this is quite ridiculous. If he can't learn simple biological facts he would have no chance of doing the work of a specialist, and it would be a sheer waste of time both on his part and of those who would have to teach him.' Gurdon can recall the words, written over 60 years ago, as if it were yesterday. He even keeps the old report pinned above the desk in his office because it amuses him. And he has every right to be amused. He has done what every schoolchild longs to do: prove his teachers spectacularly wrong.

For his experiments, Gurdon chose a different frog to study. If *Rana pipiens* is a frog supermodel, then Gurdon's frog, the African clawed frog (*Xenopus laevis*) would struggle to get a photo-shoot for a knitting catalogue. Imagine a ball of dirty brown modelling clay that's been catapulted at a wall, then had frog arms and legs stuck on. That's *Xenopus laevis*. As frogs go, this one more than deserves a kiss from a princess. But as we all know, it's not what you look like, but what you do with your life that matters, and this frog was part of arguably one of the most important pieces of biological research of all time. Gurdon chose to isolate nuclei, not from embryonic frogs as Briggs and King had done, but from tadpoles, specifically from the cells that line the tadpole's gut. The rationale was that these 'intestinal epithelial' cells lacked totipotency. These were cells that could not generate other cell types or change their own identity. Once an intestinal epithelial cell, always an intestinal epithelial cell. Researchers talk about cells such as these being 'committed', a nice analogy that somehow implies the cell has taken a vow of 'intestinal epithelialness'. It will never covet other cell types again. If the nucleus from a committed cell could direct development after being plonked into an enucleated egg,* it meant that the genes for directing development were all still there. Gurdon transferred the tadpole nucleus into enucleated frog eggs and managed to produce not one, but 10 cloned tadpoles. He proved that genetic material from mature, committed cells still contains all the information needed to make an entire organism. It was a conceptual leap forwards, and one that has spurred the fields of stem cell biology and regenerative medicine ever since. It paved the way for mammals to be cloned, for Dolly and forays into de-extinction. But it also set the stage for therapeutic cloning. Gurdon realised that if you could take DNA from someone who was ill, with Alzheimer's, say, and reprogramme it to an embryonic state through nuclear transfer, you could use the resulting embryo to harvest 'spare part' cells that could be used to treat the disease. The same cells could also be used in culture to test new medicines, speeding up the process of drug development. In 2012, Gurdon was awarded the Nobel Prize for Physiology or Medicine for his research into cellular reprogramming, and he remains an inspiration to anyone out there who, like me, suffered from crushingly low teacher expectations and perennially poor school reports. You don't have to be top of the class to do well in life.

But what of the gastric-brooding frog? Does John Gurdon's work bring the dream of resurrecting the bizarre baby-burper back to life any closer? You'd be forgiven, thanks to Gurdon's seminal frog-cloning work, for thinking that it's

---

*One that's had its nucleus removed.

easy to clone frogs. If they could do it in the fifties, surely it'd be a doddle to do decades later …

Almost Tadpoles

The first step for Michael Archer and the Lazarus Project was to find a source of gastric-brooding frog DNA. So Archer called up his friend, Michael Tyler, who had kept the amphibians in his laboratory in the seventies, to ask if he had saved any of the frogs after they died. Much to his delight, he found that he had. Decades before, Tyler had had the foresight to realise that one day, material from the gastric-brooding frog might be of interest to others. The remains of a couple of dozen individuals lay almost forgotten at the bottom of his laboratory freezer. There were whole frogs, bits of frogs, adults and juveniles; a veritable treasure trove of cells and hopefully DNA for Archer's cloning experiments.

However, the body parts were now over 30 years old and had been frozen without cryopreservatives. So when the scientists thawed a bit of tissue and tried to grow cells from it, the cells obstinately refused to play ball. They just sat there in the petri dish, floating around, long dead. When biologists cloned frogs during the fifties, their starting material was in good shape. The gastric-brooding frog tissue, in contrast, seemed of such poor quality that the Lazarus team fully expected their cloning experiments to croak. And where Briggs and King and then Gurdon had used DNA from embryonic frogs and tadpoles respectively, Archer's team was having to use DNA from much older specimens. Their cells were more mature, more specialised and arguably less able to have their biological clock 'reset' to drive embryonic development. 'It is relatively easy to obtain cloned animals starting with nuclei from embryonic cells,' says Gurdon. 'It is much more difficult to obtain normal adult animals from the nuclei of specialised cells.'

This was the plan. Drawing on the experiences of Gurdon, Briggs and King and many others, the Lazarus team planned to take the nucleus from one of their hopeless-looking cells and inject it into a live frog egg that had had its nucleus removed or inactivated. But because there were no gastric-brooding frog eggs in the freezer, the team would have to use the eggs of a living, related species for their experiments. It's another layer of complication that only lengthens the odds of success. 'No one has ever cloned a frog using the nucleus of one species in the enucleated egg of another species,' says Gurdon. On Michael Mahony's advice, they decided to plump for the Great Barred Frog (*Mixophyes fasciolatus*), a common-enough frog with a big, yolk-rich egg, which could be collected from the wild by driving around the rainforests of New South Wales at night (where Mahony is based and the cloning experiments were to be done) and trying not to run them over as they hopped across the road.

With the frogs and their eggs safely collected, they began their experiments. Very carefully, they transferred hundreds of nuclei into hundreds of recipient eggs and waited. Freshly fertilised 'regular' frog eggs take a couple of hours to start dividing, so if the cloned eggs were to do the same, the team reckoned it would probably happen on a similar timescale. But the clock ticked, hours passed and the reconstituted eggs, disappointingly, did nothing.

But then, unexpectedly, one of the reconfigured eggs did start to divide. The single cell split into two cells, which split into four cells and kept on going. The team high-fived around the laboratory. The nurturing environment of the Great Barred Frog's egg had somehow reactivated the dead nucleus and a tiny gastric-brooding frog embryo was beginning to form in front of their very eyes. 'When we saw the first, then the second, then the third cell division, we knew we were on to something,' says Mahony. The Lazarus nucleus had come back to life. But where biblical comeback king Lazarus of Bethany endured an alleged four days of death before being brought back to life by a famous bloke with a beard,* the DNA of the dead gastric-brooding frog had somehow survived 30 years in an Australian chest freezer. It was nothing short of a miracle. The team had successfully de-extincted the southern gastric-brooding frog.

Back in the petri dish, the plucky cell continued to divide until it had formed a tiny, featureless ball of a few hundred cells. That may not sound like much, but it's from such humble beginnings that all multicellular life begins. But then, just as suddenly as it had started, cell division stopped. The embryo never turned into a tadpole, much less a frog. The gastric-brooding frog was gone all over again.

Genetic tests confirmed that the early embryo had indeed contained gastric-brooding frog DNA, and that the DNA was being replicated as the cells divided. It was a genuine clone, but a clone that unfortunately never made it past the first few days of life. Still, Archer announced the team's progress at the 2013 TEDx De-extinction event to a heart-felt round of applause and a flurry of media reports. Later that year, the team bagged a place in *Time* magazine's top 25 inventions of 2013, alongside an artificial pancreas, a new atomic clock and the 'cronut', the bastard pastry child of a croissant and a doughnut.

The Lazarus frog had failed to make it past an important milestone in embryonic development known as gastrulation. A crucial period in early life, gastrulation occurs when the hollow ball of early embryonic cells starts to fold in on itself to form a layered structure known as the gastrula. These layers then go on to generate various defined tissue types such as blood vessels, bone, skin and brain. Organs and body parts can't develop until gastrulation has occurred, and the whole process is underscored by a complicated flurry of genetic and molecular activity. For some as-yet

*Not Harvard biologist George Church, the other one.

unknown reason, the Lazarus embryos stopped dividing just as they were on the brink of gastrulation.

But the team was unfazed. With every cloning attempt, they ran a detailed series of control experiments. In one, they kept everything the same, but instead of transferring a gastric-brooding frog nucleus into an enucleated egg, they injected a nucleus taken from a living Great Barred Frog. The modified egg divided for a few days, just as the cloned gastric frog embryos had done, then stopped at exactly the same point, just before gastrulation. It sounds like bad news, except that it isn't. Because embryos from both living and dead donor nuclei fail at the same point, the control experiment suggests that it's probably not the dead gastric-brooding frog nuclei that are the problem. More likely, it's the cloning process itself. So if the cloning process can be improved, perhaps the Lazarus embryos can be nurtured beyond gastrulation, into tadpoles and then into frogs.

The Lazarus scientists believe that the gastric-brooding frog's DNA still contains the information needed for making a baby frog, it's just a matter of getting the egg to reactivate these genetic signals. 'I think it's the egg that is the brick wall,' says Mahony. Just as egg and sperm are vital for normal fertilisation, so too the egg plays a vital role when life is created through cloning. Its cytoplasm (that's pretty much everything inside the egg minus the nucleus) contains an as-yet uncharacterised cocktail of molecules that can physically reconfigure and alter the activity of DNA. These molecules can reorganise, repackage and reactivate DNA, somehow switching on the signals that guide embryonic development. But if the egg becomes damaged, the signals can fail. Embryonic development may not happen at all, or if it does, it may not last that long. Although the eggs of the Great Barred Frog 'look' good, perhaps they are the reason that Lazarus embryos have yet to turn into anything resembling a tadpole. It's also hard to remove every last scrap of Great Barred Frog nuclear DNA from the egg, another hurdle the team is working hard to overcome.

They're trying to turn things around by using only the best eggs for their cloning experiments. And it's here, speaking with another member of the team, Andrew French, that I realise the level of detail, dedication and doggedness required to take the Lazarus embryo to the next level. Let me explain. It'll give you an idea of just how hard these guys are working. To get the best eggs requires the best frogs, but the Great Barred Frog is a seasonal breeder that only lays in the summer months between September and February. It's on these warm, wet nights that one of the team hops into a four-wheel drive and cruises around the forests of New South Wales, collecting females as they crawl out of the leaf litter and cross the road to hop down to the creek to breed. It takes a few weeks of sleepless nights to collect the few dozen females needed for a round of cloning experiments, and the ladies are then chauffeur-driven back to the lab where they are treated like frog royalty. 'We need to get them in tip-top shape,' says French. So they are fed titbits of chicken and live free-range crickets. Just as with human IVF, the frogs then receive hormone treatment to stimulate ovulation. But you can't buy frog hormones over the counter, so instead the team has to 'make' and purify the hormones themselves, from the pituitary glands of cane toads. The hormones are then injected into the females at roughly the same time so that they all, in theory at least, ovulate their 500 or so eggs on the same day – the designated day for the cloning experiments … except that it never really pans out that way. For every five or so hormone-injected females, only one or two ovulate when they are meant to. French can tell when this is happening because ovulating females frantically wiggle their back legs to and fro in an almost contraction-like fashion. The eggs need to be used when they are as fresh as possible, so if the female doesn't lay them straight away, French will delicately massage her abdomen until she does. But the freshly laid frog eggs are still not ready for cloning experiments. Ask any five-year-old and they will tell you that frog eggs sit inside a protective layer of jelly. It's this gelatinous mass that is the nemesis of frog-cloners. There is, in fact, not one but multiple layers of jelly surrounding the egg, and in texture it's like the silicone putty used to seal windows. Peel one layer away and another seems to miraculously spring up in its place. Sperm can somehow manage to get through these multiple membrane layers, but the fine glass needles of cloners cannot. The needle either bounces off or shatters. These are slimy, difficult eggs to work with. It's incredibly frustrating, so the team is refining methods to de-jelly their frog eggs. It's a complex procedure that involves chemicals, UV light and physically teasing the membranes off with fine forceps. 'We are having to re-invent Gurdon's protocols from the 1950s,' says French. The hope is that by collecting and treating the eggs with extreme care and respect, by paying attention to every single component of the protocol leading up to nuclear transfer, the eggs will be in top shape, and able to reactivate the dead, DNA-containing nuclei from gastric-brooding frogs. And it looks as if their hard work is finally paying off.

'Every time we re-run the experiments we get better results,' says French. For every 100 or so nuclear transfer experiments done – where the nucleus from a dead gastric-brooding frog cell is injected into a de-jellied, enucleated live frog egg – around seven or eight live Lazarus embryos are created. Most fail before gastrulation but every now and then, once in a while, when there's a blue moon in the sky and the planets align, the odd cloned embryo keeps on dividing. The team has watched with great excitement as a small number of amorphous cloned Lazarus embryos begin to develop recognisable tadpole-like features, in the shape of a small bumpy ridge destined to become the animal's backbone. 'They're almost tadpoles,' says French. Little by little, as the experimental protocols creep closer to the optimum, the embryonic frogs are starting to live a little bit longer, to develop a little bit further and to become more tadpole-like. The frustration for now, however, is that the 'almost tadpoles' remain just that: 'almost'. They don't turn into tadpoles, much less froglets or frogs. Embryonic development grinds to a halt after the primitive backbone is made, and the embryos all die.

But I'm not disheartened. I've spoken to these scientists and I know they're not going to give up easily. They have optimism, team spirit and a never-ending list of experimental variables they will continue to tweak until they get their procedures 'just so'. Next step, to take a nucleus from a living cloned embryonic gastric-brooding embryo (before it stops dividing) and use that for cloning. 'We think that re-cloning might be the best option,' says French. By cloning a clone they hope to push subsequent embryonic development all the way and bring the gastric-brooding frog back from the dead.

When British scientists Patrick Steptoe and Robert Edwards first tried to create human embryos in a dish back in the late sixties, they found it immensely difficult. Their attempts to artificially meld human sperm and human egg met with failure time and time again. Indeed, it took the pioneers of *in vitro* fertilisation (IVF) eight years from when they first witnessed an IVF egg dividing in a dish to the birth of the world's first test-tube baby, Louise Brown, on 25 July 1978. It's hugely unrealistic to expect experiments of this nature to work anything like first time. 'Ultimately you need all the conditions to be right on the day … and then it's a numbers game,' says French. Transfer enough donor nuclei into enough eggs, continue to learn from your experiments and refine your protocols, and eventually the gastric-brooding frog will rise Lazarus-like, not from a stony grave, but from the bottom of a chest freezer in a laboratory in Adelaide.

'Watch this space,' says Michael Archer. 'I think we're going to have this frog hopping glad to be back in the world again.' If I were to choose one animal from the whole of the history of time to bring back to life again, it might well be the gastric-brooding frog.

CHAPTER SIX

King of Rock 'n' Roll

When it finally arrived, I was more than a little disappointed. Some time earlier, I had bought some of Elvis Presley's hair on eBay and, like a child waiting for Christmas, had then had to endure weeks of over-excitement and speculation while I waited for my purchase to turn up. I was expecting something as iconic as the King himself. The package, I told myself, would at the very least be made of gold lamé and studded in rhinestones. It would arrive with great ceremony, serenaded by its own backing band, willed on by the frenzied screams of female fans who would camp outside my house and faint when the parcel arrived. But when the postman finally delivered my precious cargo, it was without fanfare. He didn't even ring the bell. He just popped the plain brown envelope through the letterbox and walked away.

I picked up the drab, un-Elvis-y anticlimax and turned it over in my hands. Save for the Memphis postmark smudged across the top right-hand corner, there was little hint as to the import's legendary contents. Perhaps things would be more promising on the inside, I told myself. Brimming with nervous energy – all shook up, even – I carefully opened the envelope and teased out the contents. But the luxuriant quiff of my dreams was nowhere to be seen. Instead, I found myself staring at an A4-sized 'Certificate of Authenticity'. 'You are now the proud owner of a genuine strand of Elvis Presley's hair,' the certificate crowed. But where was the hair? Perhaps they had forgotten to include it, or maybe it was being sent separately under armed guard from Graceland. I scoured the shoddily photocopied piece of paper for a clue. And it was only then, under the full scrutiny of my most intense Paddington Bear stare that I noticed a blob of dried glue at the bottom of the page. It was pea-sized and transparent, easy to miss, but inside it, like an insect trapped in amber, was a short strand of hair. I stared at it, unable to believe what my eyes were seeing, incapable of processing what I held in my hands. Words actually failed me. Not only was the 'strand' minuscule – smaller, even, than a single eyelash from my face – it was also … I can't bring myself to say it … ginger. Elvis Presley, if the certificate was to be believed, was strawberry blonde.

As the penny dropped I felt so lonesome I could cry. I had spent $14.90 (plus postage and packing) and endured a longer wait than the King's fans had for his comeback tour, only to receive what appeared to be Ronald McDonald's nasal clippings. So I did the only thing I could think of. I sealed the certificate back up in its original envelope, found a thick black marker pen and wrote upon it, 'RETURN TO SENDER', adding that the address was unknown. The irony, I was sure, would be lost on them, but it put a smile back on my face. Then bright and early next morning, I sent my letter back.

I'm not sure, to this day, exactly why I decided to buy a strand of Elvis Presley's hair. It's true, I've been a fan most of my life. I have an Elvis clock in my study, an Elvis jumpsuit in my wardrobe and a be-quiffed hamster that sings 'Rock-a-hula'. When I got married, my lovely mum walked me down the aisle to the sounds of 'Burning Love'.* But I'm not obsessed. I just like his music and the kitsch paraphernalia that goes with it. So when I stumbled across a strand of his hair during my daily trawl of Elvis eBay ephemera, I simply had to have it. Hair, I knew, is an excellent source of DNA, so the purchase offered the chance to buy a piece of genetic history, the very DNA that helped to shape one of the most talented and beautiful men on the planet. I bought a strand of Presley's hair because it seemed the ultimate Elvis trinket. I bought it out of academic curiosity. I bought it because it got me thinking ...

... if you could retrieve Presley's genome from a lock of hair or some other source, could you somehow use it to bring back the King? Could you 'de-extinct' Elvis Presley? And if you did, how similar or different would the new Elvis be from the original? Did Elvis have sexy DNA? Was there some part of his genome that endowed him with his famous pouty sneer or extraordinarily agile hips? How about his predilection for fried peanut butter and banana sandwiches, or his penchant for blue suede shoes? Would we bring him back only to condemn him to a second untimely toilet-related death?

Now, at this point, there will, I'm sure, be those among you wanting to gnaw off their own legs in agitation. *'There's no point cloning pop stars,'* I hear you cry. *'Simon Cowell has been doing that for years and they're all rubbish ... And besides, Elvis is dead, not extinct, so you can't talk about de-extincting him ... OK?'*

---

*'Burning Love' was my request. My husband wanted 'Heaven Knows I'm Miserable Now' by the Smiths.

To which I reply, fair point about Simon Cowell. However, human beings may not be extinct … yet. But our time, inevitably, will come to pass. So when, not if, we become extinct, will we leave behind the instruction manual and the technology for de-extincting ourselves? People, myself included, have mused on the possibility of de-extincting ancient humans, but few have dared to draw the thought experiment to its logical conclusion: whether or not we ourselves could be de-extincted. We'd be wise, at least, to think about it. And if we are going to think about it, we might as well make our test subject easy on the eye … and on the ear. A discussion about de-extincting *Homo sapiens* is not foolish. It's prescient. And to anyone who finds themselves shouting, *'Sexy DNA? Are you mad, woman? There's no such thing as "sexy DNA". The phenotype of any human being is a complex interaction of genetic, epigenetic, environmental, not to mention stochastic factors,'* I say, calm down, a little *less* conversation. I'll get to all that. But for now, let's rock 'n' roll.

In the Ghetto

Elvis Aaron Presley – the real deal – was born around noon on 8 January 1935, in Tupelo, Mississippi. His mother, Gladys Smith Presley, a poor cotton picker, gave birth to twins. But Elvis's brother, Jesse Garon, was stillborn. Elvis's father, Vernon Elvis Presley, drifted from job to job and spent time in jail, and Elvis grew up an only child in the family's modest two-room wooden shack. He had a very musical childhood. In the evening, he would sit on the porch with his family and sing. At church, he would squirm out of his mother's arms so he could join in with the choir. He listened to Pentecostal church music, musicians singing the blues and the *Grand Ole Opry*, a weekly country-music concert that was broadcast on the radio. Then, aged 10 or 11, his mum took him to the local hardware store and bought him his first guitar.

In 1948 the family moved to Memphis, Tennessee, where Elvis restyled his hair into a ducktail-do to hide his teenage acne. He heard jazz in the legendary Beale Street, and then one day, the story goes, he drove past a sign that changed his life. It was an advert for the Memphis Recording Service, a 'make your own' record set-up charging four dollars for two songs. Elvis decided to record a couple for his mum, but the studio's owner, Sam Philips, was so impressed by his vocal range that he signed Presley up to his record label, Sun Records. In 1954, Elvis released his first single, 'That's All Right'. According to *Rolling Stone* magazine it was the first rock 'n' roll record ever made. A year later, fans rioted outside his concert in Jacksonville, Florida, when Elvis finished his show by cheekily announcing, 'Girls, I'll see you backstage.'

At 19 years old, Elvis was already well on his way to becoming an international sensation. With a sound and style like nothing before, a dangerously mobile pelvis and an electrifying stage presence, Elvis ushered in a new era of US music and popular culture. His gyrating hips were deemed so sexy that certain TV programmes refused to broadcast the bottom half of his body. Then in 1956, with a new record deal under his belt, he released 'Heartbreak Hotel'. It was the first record to simultaneously reach number one in three different charts: the *Billboard* Country and Western chart, the Rhythm and Blues chart and the Pop chart. After that, the movies beckoned. Elvis released a string of 33 films and moved into the ultimate man cave, Graceland. By the time he was drafted into the US Army he had released 14 consecutive million-selling records. In 1967 he married Priscilla, and a year later returned to TV with his electrifying, sultry comeback special called simply *Elvis*. Black leather has never looked so good.

In the years that followed, he experienced some of his highest and lowest moments. He released 'In the Ghetto' and 'Suspicious Minds', but his health was starting to fail. The pounds were piling on and Elvis became addicted to prescription drugs. In 1973, amidst his busiest touring schedule ever, he divorced from Priscilla and overdosed twice on barbiturates. He suffered from migraines and glaucoma, slurred his way through his performances, and towards the end of his life had to be physically guided off stage by his minders. In the first eight months of 1977, he was prescribed more than 10,000 doses of sedatives, amphetamines and narcotics. Then, on 17 August 1977, his girlfriend Ginger Alden found him dead on the bathroom floor of Graceland. He was just 42 years old. President Jimmy Carter issued a statement crediting Elvis with having 'permanently changed the face of American popular culture.' Muhammad Ali called him the 'sweetest, most humble and nicest man you'd want to know' … but could we ever bring him back?

Americans for Cloning Elvis

Once upon a time, in the mid-to-late nineties, when mobile phones weren't smart and the internet had more hand-drawn porn than crap cat pictures, there was a website called 'Americans for Cloning Elvis'.* Inspired by the birth of Dolly the sheep, it featured a grainy image of Elvis gyrating his hips awkwardly as if locked in some eternal digital spasm. The title, 'Americans for Cloning Elvis', was written in a retro, pixelated font, and underneath it was a call to arms. The website implored like-minded Elvis fans to sign a petition asking scientists to bring the rock legend back to life. It read:

> We the undersigned, in our enduring love for Elvis, implore all those involved in cloning to hear our plea …

*The website still exists and can now be found at americansforcloningelvis.bobmeyer99.com.

> The technology is here, and this petition is a testament to our will.

There are Americans, it seems, who would genuinely like to see Elvis cloned; either that or they've too much time on their hands and like signing petitions. Today, the website is still live and largely unchanged, and more than 3,000 people have added their name to the cause. How serious they are is up for grabs. They don't, I'm fairly certain, hang out in secret laboratories practising nuclear transfer on embryos with quiffs, but they have at least thought about cloning the King. But what would it take? Where to start?

The plan would involve finding a source of Elvis's DNA, decoding his genome then editing the parts of it that were unique to him into a regular human cell. The DNA-containing nucleus of this cell could then be used for cloning to create a baby that would effectively be Presley's identical twin. At face value it's not so different to the method being used by George Church to make a hairy elephant, but this is Elvis we're talking about. Where mammoth DNA can be found with relative ease in the cells of thawing Arctic carcasses, there was only ever one Elvis Presley, and he's currently pushing up the daisies at Graceland.

But you can, as my own experience has shown, buy snippets of famous people's hair on the internet and through private dealers.* A few clicks on Google turn up locks belonging to the likes of Princess Diana, Abraham Lincoln

---

*On a more macabre and unpalatable note, body parts from dead celebrities are also up for grabs. Napoleon's penis, for example, is said to be in the possession of one Evan Lattimer, daughter of a New Jersey urologist who bought the item at a Paris auction in 1977. Apparently removed at autopsy by Napoleon's doctor, the shrunken member is said to be around 4cm (1.5in) long. Josephine probably wasn't that disappointed when he said 'not tonight'.

and Michael Jackson.* Just as authenticity varies, so does price. Although I may have parted with the best part of $15 for something that looked suspiciously like a ginger pubic hair, a genuine tuft of Che Guevara's hair is reported to have sold for $119,500. And when Britney Spears had her head shaved in 2007, her hairdresser tried to auction the locks online with a starting price of $1 million.†

But if you want to find anything to do with Elvis, there's really only one place to go: the Loudermilk Boarding House Museum in Cornelia, Habersham County, Georgia. From the outside it looks unassuming, a pretty timber-built structure with twin gables and a big white veranda. But on the inside is the quirkiest, craziest collection of Elvis memorabilia that there is, comprising Joni Mabe's 'Travelling Panoramic Encyclopaedia of Everything Elvis'. Mabe, a talented artist and self-proclaimed 'Queen of the King', has been hoarding Elvis knick-knacks ever since the day the King 'left the building'. She has five rooms of glitzy collectables stacked from sequinned floor to satin ceiling. Her collection, which toured the world for 14 years, is a haven for the kitsch and the retro. Aside from the more obvious costumes, tabloid clippings and bubble-gum cards, there are Presley prayer rugs, shoestrings and toenail clippers. There's even the toe tag that Elvis wore as he lay in his Memphis funeral home.

Amidst it all, however, are three possible sources of Elvis DNA.

The first is not just a strand, but an entire lock of Presley's hair. Dyed black in alleged homage to fifties movie star Tony Curtis, Presley's pomade-laden quiff required some serious maintenance at the hands of his personal barber Homer Gilland, who squirrelled away and then sold much of the discarded hair. Mabe bought her sample from him directly, so is convinced of its authenticity.

---

*Presumably pre-Pepsi commercial.

†But no one bought it.

Although the hair on our heads is dead, the hair cells at the base of each strand are very much alive. These divide to generate new hair cells, which then migrate into the hair shaft and become infilled with keratin, the structural protein that makes hair hard. As the cells become keratinised, their DNA starts to break down, but fragments can sometimes still be retrieved many thousands of years later. Mitochondrial DNA has been recovered from the fur of woolly rhinos and mammoths and, in 2010, scientists recovered the full nuclear genome of a 4,500-year-old mummified Inuit ... all from his hair. It's reasonable therefore to assume that genuine Elvis hair, just decades old, could be a source of his DNA. Nor should Presley's proclivity for hair dye be a problem. When researchers extract DNA from hair they bleach it first then rinse it with alcohol. The step is not cosmetic; rather, it removes any contaminating DNA present on the outside of the strand. In Presley's case, this means that any DNA recovered would come from the man himself, and not his hairdresser or anyone else.

A second potential source of DNA comes from another keratinised structure, this time one that Mabe collected herself when she visited Graceland in 1983. Passing through the dimly lit Jungle Room, Presley's green-carpeted 'man cave' of a den, Mabe became overwhelmed with the urge to literally feel what Elvis had felt. As her tour group moved on, she lingered behind to skim her fingers across the shag pile, only to feel them catch on what she thought was a wayward sequin. It was only when she examined her treasure outside in the broad daylight, that she realised she had found a toenail clipping. 'It's either from a second toe, or it's a piece of the big toenail,' she says. The sign of a true Elvis obsessive, Mabe has kept the toenail for more than 30 years. Just like hair, nails are also a good source of DNA, but the problem here, is that no one can be sure who the toenail belonged to. A lot of people partied in the Jungle Room, and even more have passed through it since Elvis shuffled off this mortal coil. 'I call it the Maybe Elvis toenail,' she says, in her glorious Southern drawl, 'because maybe it's from Elvis, and maybe it isn't.' Maybe the toenail *is* from Elvis, but maybe it's from the guy who laid the carpet. If we're talking about bringing back the King, 'maybe' isn't good enough, which leaves option three.

Study photographs of Elvis pre-1958, and you'll notice he has a wart on the back of his right wrist. Subsequent pictures clearly show the dark blob has gone. It was removed by Presley's personal doctor, who then popped it into a test tube of formaldehyde and kept it for three decades until Mabe bought it from him for an undisclosed amount. The wart now has pride of place in Mabe's museum, where it remains one of her most popular exhibits. She even sells T-shirts with the logo 'The King is Gone but the Wart Lives On'. The wart is still in its original container, but the tube is now propped up on a bed of red silk in an old cigar box. 'It's as big as a black-eyed pea, but it's real pus-y,' says Mabe who seems to relish telling me the gory details. But could Elvis be de-extincted from it?

DNA can readily be retrieved from formaldehyde-preserved tissue but warts don't make for the best starting material. Put aside for one moment the 'yuk' factor and the psychological trauma of finding you've been created from an unsightly blemish, there are other problems too.

Warts occur when a skin cell becomes infected with the human papilloma virus (HPV). Inside the cell, the viral DNA hijacks the host's DNA, making copies of itself that can then invade and infect neighbouring cells. Infected cells divide more rapidly than non-infected cells, causing the wart to grow. Sometimes this abnormal growth gets out of hand, and an originally benign wart can become cancerous. It's for this reason that Presley would have had his wart removed.

Were someone to extract the DNA from an HPV-infected cell they'd find a mix of human and viral genetic material.

The human genome, at three billion base pairs, dwarfs the HPV genome, which consists of a runty 8,000. But size, as Napoleon undoubtedly said to Josephine, isn't everything. It's what that tiny genome could do that would be troublesome. If the Elvis/virus DNA mix was inserted into an empty egg and used for cloning, then every cell in the developing embryo runs the risk of having both Elvis DNA and virus DNA. Given HPV's reputation for causing abnormal cell growth and cancer, this could have unpredictable and potentially dangerous consequences. The developing human embryo is a miracle of finely tuned genetic activity. Chucking an extra load of 'alien' DNA into the mix would be like throwing a spanner into a car engine. If you're lucky, nothing much happens. If you're unlucky, the engine dies. Warts are not a suitable source of DNA for cloning.

So it looks like Elvis's iconic coiffure, or part of it, is the best place to begin our quest to bring back the King. After the best part of half a century, Presley's genome would be expected to have broken into millions of shorter fragments, but this shouldn't be a problem. In 2014, scientists working for the Channel 4 documentary *Dead Famous DNA* took a sample of Elvis Presley's hair and managed to extract DNA from it. They didn't try to decode the entire genome; rather they read the sequence of some of the snippets that they recovered (of which more later). But if they or anyone were to sequence enough fragments enough times, eventually the whole genome could be decoded and reassembled inside a computer. Stored in digital form, it would be the longest recording Presley ever made.

But it wouldn't be complete. In 2004, researchers announced that they had finished sequencing the human genome. It was, the then-US president Bill Clinton said, 'the most wondrous map ever produced by human kind.' One of the most significant scientific milestones of all time, people compared it to the invention of the wheel or the splitting of the atom. The culmination of more than a decade's work by thousands of scientists across the globe, it promised unparalleled insights into human development and disease, and to pave the way for the development of new therapies. But it wasn't complete. When they set out to decode the human genome, researchers decided to focus on only the most active, gene-dense regions, the so-called euchromatin, because, they reasoned, these were likely to be the most important bits. The remainder of the genome, a small gene-light fraction of the total DNA known as the heterochromatin, was thought less important. It was also fiendishly tricky to decipher, so researchers decided to leave it well alone. The result: when the human genome sequence was finally 'finished' and published in the journal *Nature*, the heterochromatin was missing, and the sequence they had for the euchromatin was full of holes, around 340 of them. According to one of the partners in the project, the National Human Genome Research Institute, the human genome was as complete as it could be 'within the limits of today's technology'. It's like saying my family holiday to France was complete when we got to Dover and missed the ferry – the technology required to get us across the Channel was simply not there at the time.

Since then, DNA sequencing has improved. Its cost has dropped and the technology itself has become widely available. But even the most thoroughly sequenced genomes are still incomplete. Although researchers have made inroads into sequencing the difficult-to-read heterochromatin, the best-sequenced human genomes are still only around 95 per cent complete. The problem is that parts of the genome are full of repetitive, stuttering snippets of code. Imagine buying a thousand copies of Richard Dawkins's *The God Delusion*,* then putting them through a shredder† and trying to reassemble them. The word 'God' will be repeated many times, but it's hard to know which 'God'

*Just to wind up the creationists.

†Just to wind up Richard Dawkins.

went where. Just like the human genome, it's hard to reassemble these tricky bits into the correct order.

Also, there are mistakes. We're only human after all. The official human genome sequence had an overall error rate of less than one error per 100,000 base pairs. That means that, although around 2,999,970,000 letters of the code are correct, 30,000 or so of them will be wrong. It's for this reason that geneticists like to check and double check their sequencing data. They call it 'coverage'. If a genome has 30x coverage – pretty much the best that there is – it means that most parts of the genome will have been double checked 30 times or more. But other parts of the genome will have been checked fewer times than this, and some regions will not have been read at all. A completely complete human genome, it seems, is wishful thinking.

If we were trying to clone a generic human, then these inconsistencies and omissions would probably count for little; most single letter changes to the genome have no effect, and gaps in one genome could be filled in with appropriate sequences from another. But because we're trying to clone a specific human it becomes a potential problem. The plan, remember, is to sequence Presley's genome then edit the parts of it that were unique to him into a regular human cell. But if these unique Elvis sequences lie in parts of the genome that either can't be read or have been read incorrectly, we run the risk of missing them altogether.

Essence d'Elvis

That Elvis was one of a kind is a claim without doubt, but his DNA was reassuringly mundane. Scientists have sequenced bits of celebrity genomes, including DNA sequences from Charles Darwin, Isaac Newton and Elvis Presley, and found them to be made of the same four chemical letters – A, C, G and T – that make up all our genomes. They are not bigger, studded with stars or decorated with dollar bills. But they are all unique.

In 2012, researchers announced that they had sequenced and compared the genomes of 1,092 different people from around the globe. These weren't celebrities, just everyday people, and the goal of the study, called the '1000 Genomes Project', was to understand what makes people's genomes different. What they found is if you compare any one person's genome with that of a non-relative, there are around three million differences, or 'variants'. That means there are around three million genetic differences between you and me, between me and Elvis, and between Elvis Presley and any non-related person on the planet. Our genetic sequences differ at about one place in every 1,000 or so letters. Most of the differences are small. An A may be substituted for a T. Perhaps a G is left out somewhere. Maybe there's an extra C hiding away somewhere else. If the human genome were a book, then everybody's book would contain the same chapters, paragraphs, sentences and words, arranged pretty much in the same order. But my book might be missing an apostrophe on page 521,* while on the next page your book might say the word 'pants' instead of 'parts'. Every now and then, however, bigger changes crop up. Whole paragraphs might be duplicated, and words might be repeated. A person with Down's syndrome, for example, has an entire extra copy of chromosome 21, while a person with Huntington's disease, a brutal and incurable neurodegenerative condition, has many extra copies of the sequence 'CAG' hiding inside a particular gene.

The differences occur because when regular cells divide – to create new cells, to build body parts, to help injuries heal – they don't always do it perfectly. In an ideal world, when one cell splits into two, the genome is duplicated

---

*Unforgiveable!

then divvied up identically. Each 'daughter' cell receives an exact copy of the original 'parent' genome. But the copying mechanism isn't perfect, and mistakes, or 'mutations', creep in. When a mutation occurs in a sperm or an egg cell it is passed on to the next generation. The result, your genome, contains about 60 'new' mutations that you inherited from your parents and many older mutations from further back in your family tree.* In addition, our genome is constantly under assault from DNA-damaging radiation, pollutants, chemicals and the like; and although most of this damage is patched up, we continue to accrue changes to our genetic code with every day we are alive.

With seven billion people on the planet and counting, the individual genetic variants that you carry are unlikely to be unique to you. I may have a 'C' instead of a 'T' towards the end of chromosome 2, but you might do too. But you won't have all of the other 2,999,999 genetic variants that make me me. 'Almost none of your variation is unique to you,' says geneticist Gilean McVean from the University of Oxford, one of the leaders of the 1000 Genomes Project. 'What is unique about you is the combination of the variants that you have.' Hidden somewhere amidst the three million genetic variants lurking in Elvis Presley's genome is the genetic essence of Elvis, the collective spattering of idiosyncratic nucleotides that helped to make him the person that he was.

In the last few years, geneticists have begun to link combinations of these variants to all sorts of diverse characteristics, such as height, obesity and mathematical ability. The combined action of many, many variants scattered across our genome are now thought to influence

---

*This is largely the fault of the men in your family. Men make lots of sperm, while women make comparatively few eggs. So the cells that divide to produce sperm do so much more frequently than the cells that divide to make eggs. Men make more mistakes. It's scientifically proven.

everything from the way we grow to the way we behave to the diseases we develop. 'It's like a background genetic effect,' says McVean. 'It could very subtly influence everything from the shape of your nose to whether or not you like gherkins.' It's these tiny differences, these individual foibles and features, that in part help to make us who we are. My personal collection of genetic variants could help explain why my nose looks like a blob of pasty-coloured modelling clay, and why, while I welcome the gherkin, I eschew the cornichon. Even though on their own, the vast majority of Presley's genetic variants will have had little or no effect, put them altogether and there you have it: the genetic essence of Elvis. 'The constellation of factors that make up Elvis are a result of many, many weak influences across the genome,' says McVean.

That means if we want to create a genetic replica of Elvis we can't skimp on the detail. If there are three million genetic differences between Presley's genome and that of a mere mortal, then to de-extinct the King, we're going to have to edit each and every single one of them into a human cell. It's too big a job for that darling of the gene-editing world, CRISPR, so geneticists would instead have to turn to CRISPR's big brother, MAGE, which can bulk-edit multiple changes into a genome simultaneously (see Chapter 2). The Presleyfied genome would then be injected into a human egg that had had its own nucleus removed, and that cell would then be coaxed to start dividing in a dish.* A few days later, a healthy looking embryo would then be transferred into the uterus of a surrogate mother, who would then be advised to stock up on DVD box sets and put her feet up. All going well, nine months later, a healthy baby boy – let's call him GElvis

*'Presleyfication' is a sadly under-used piece of made-up jargon. It refers to the act or process of modifying a human genome so that it becomes the same as Elvis Presley's.

(short for Genetically engineered Elvis*) would be born. But how much like Elvis would he be? To what extent would his carefully crafted DNA influence the way he turned out?

Seeing Double

We know that Elvis was born a twin, and although it's unknown whether his stillborn brother was identical or non-identical to him, two things are certain. First, twins of any kind are incredibly special. I should know; I have twins of my own. And second, through the scientific study of twins we can glean insight not just into how GElvis might turn out, but how we all end up the way we do.

With identical DNA, or near as damn it, GElvis would effectively be Elvis's identical twin. And as anyone who has watched *The Shining* will tell you, identical twins can sometimes be spookily similar. Not only do they look alike, they sometimes share the same mannerisms, interests and habits. For the most part we presume that these similarities are down to the fact that not only do they have the same genes, but they also grow up in the same environment. They have the same parents, live in the same household and attend the same school. Much to my own twins' disgust, they also share the same bedroom and, more often than not, the same mismatched socks.

But sometimes, identical twins that grow up separately can also end up alike. In 1979, US psychologist Thomas J. Bouchard came across the Jim-twins. Adopted by different families when they were just four months old, the twins grew up unaware of one another's existence. Then Bouchard reunited them when they were 39 to discover that not only did the two brothers look alike, they shared a

*Just so we're all reading from the same page, let's pronounce the 'G' in 'GElvis' softly, like the 'G' in 'gene'.

whole list of unexpected similarities. Both bit their nails, folded their legs the same way and dangled keys from their belts. Both were married to women called Betty, divorced from women called Linda and had owned a pet dog called Toy. One named his first son James Alan Lewis, the other James Allen Springer. They both drank Miller Lite, smoked Salem cigarettes and holidayed on the same beach in Florida. And to top it all, they both drove the same colour and make of Chevy.

All in all, as part of the now classic Minnesota Study of Twins Reared Apart, Bouchard studied the fates of more than a hundred twins or triplets brought up separately. Overall, the study revealed that an identical twin reared away from his or her co-twin had about an equal chance of being similar to their sibling in terms of personality, interests and attitudes as one who had been reared with their twin sibling. The similarities between the Jims, the study suggested, were down to their genes. Put simply, the environment didn't seem to make much difference at all. Going by this example, it wouldn't matter if Elvis and GElvis were born 40 years apart, one into rags in a Tupelo shack, the other into riches in a high-tech hospital; their different environments and upbringings would count for little. Perhaps GElvis would also marry a Priscilla and have his own Lisa Marie. He might live in a big flashy pad in Memphis and even be prepared to fly 1,000 miles for a sandwich.*

But it is, of course, more complicated than that. If the environment didn't matter then factors such as parenting, education and diet would count for little … which would be nonsense and take away the God-given right of teenagers to blame their parents for *everything.* When people looked at the

*On 1 February 1976, Elvis famously flew round-trip Graceland to Denver in his private jet so he could buy an 8,000 calorie 'Fool's Gold' sandwich made from a hollowed-out loaf filled with a jar of peanut butter, a jar of jam and a pound of bacon.

Jim twins more closely, they realised that they were actually far more different than was initially assumed. One was a talker, the other a writer. One sported a Beatles-style mop top, the other a Robert de Niro-style do. Much to the probable alarm of one Jim's second and current wife, his brother had married a third time. The Jim-twins were such a scoop that the media downplayed their differences and instead focused on their 'spooky' similarities. Where the men lived, there were loads of guys that drank Miller Lite, smoked Salems and drove Chevys. The Jim-twins habits weren't eerily similar; most of them were mundanely routine.

So just how different are identical twins? We know that identical twins grow into very different adults. They may look similar but they have their own distinct personalities, and physical idiosyncrasies that become magnified with age. They have different preferences and habits, and often develop different diseases. Just like the rest of us, they are a product of genes *and* the environment, but what's up for grabs is the relative importance of each of these factors. To what extent are particular characteristics, such as, say, burger-eating or musical talent, influenced by our genes and how much are they influenced by the environment? It's a conundrum that scientists have been trying to solve ever since Darwin's cousin, Francis Galton, first started looking into it over a century ago. They call it the 'nature versus nurture debate' and twins are at the heart of it. In the last 50 years, more than 14 million twin pairs have contributed to thousands of studies trying to tease apart the relative roles of DNA and upbringing in everything from personality to piles, gun-ownership to gout.

The rationale is this. Identical twins, formed when a single fertilised egg splits in two, share virtually all of their genes. Non-identical twins, created when separate sperm fertilise separate eggs, share around half. Twins tend to grow up together, so share the same environment. So if a characteristic that you're interested in, lip-curling, for example, is similar in both types of twin, then genetics can't play much of a role. The environment is more important. But if identical twins are more similar than non-identical twins at, say, hip-swivelling, then it must be down to their genes. All in all, more than 2,700 twin studies have assessed more than 17,000 different characteristics, and the results are quite remarkable.

It turns out that pretty much whatever feature you decide to look at, the test results of identical twins are more similar than those of non-identical twins. Everything, it seems, has a genetic component. That certain diseases or physical traits tend to run in families comes as no big surprise, but twin studies also reveal that our DNA influences intelligence, obesity and addiction. But the studies go further still. Bizarre though it may seem, twin studies also hint that genetics sways certain life events and lifestyle choices. Your DNA influences whether or not you will fly or flunk at school, marry or divorce, end up a billionaire or go bankrupt. It even plays a role not just in musical ability, but in how much practice you put in.

But just because everything has a genetic component it doesn't mean that genes control our destiny. Twin studies also reveal that, while almost everything is influenced by our DNA, nothing is 100 per cent determined by it. Whatever characteristic you choose to look at, the environment always plays a role too. Nature and nurture operate in a 'pushmi-pullyu' sort of way. Some features are more influenced by genetics, others by the environment.

By studying twins, researchers can estimate the relative influence of genetics versus the environment. They call it heritability: the proportion of observed variation in a particular characteristic that can be attributed to genetic rather than environmental factors. Height and eye colour, for example, are strongly influenced by genetics, as are intelligence and obesity, while addiction, migraine, life events and my own Achilles heel, the hangover, are more influenced by non-genetic factors, like how many tequila shots your 'best friend' buys you.

We know that Elvis divorced Priscilla, that he became overweight, suffered from migraines and took too many prescription drugs. But from twin studies we should surmise not that GElvis would end up the same, rather he would be more or less likely to share certain characteristics. From estimates of heritability, we can conclude that GElvis and Elvis (and indeed all identical twins) would be more likely to be similar in terms of appearance and jumpsuit size, and more likely to be different in terms of headaches, drug habits and marital status. But it doesn't mean that GElvis *will* become obese or that he *won't* divorce. Obesity, for example, may well be one of the strongest genetically influenced characteristics there is, but that doesn't mean the environment, or, more specifically, the number of burgers in it, doesn't play an important role too.

Geneticists suspect that this pervasive influence of genetics is caused by the interaction of many genes, which on their own have little effect. Despite what you might have read in the media, there is no 'gene for' heart disease or happiness, any more than there is a 'gene for' having a quiff or being sexy. With the occasional exception, there are no 'genes for' anything specific.* What there are, however, are our unique three million variants; the bits of your genetic code that make it distinct from mine, that subtly influence whether or not you will develop a particular characteristic or disease. When scientists studied snippets of Presley's DNA for the documentary *Dead Famous DNA* they found that he carried a number of these disease-linked variants. Presley's were related to obesity, migraine and glaucoma, conditions he is known to have suffered from. But it wasn't,

*There are several thousand single-gene diseases, so called because they are caused by defects in single genes. So there is a 'gene for' Huntington's disease and cystic fibrosis. But the vast majority of human diseases and characteristics are complex, caused by poorly understood combinations of multiple genes and other factors.

as the show pontificated, the smoking gun explaining Presley's ill health. Data from the 1000 Genomes Project suggests that we all have around 5,000 or so of these disease-linked variants lurking somewhere in our genome. I have them. You have them. But we don't all go on to develop the diseases they are linked to. Apart from the odd exception, our genes and genetic variants don't predestine us to ill health, or anything else for that matter; they just slightly skew the odds that we will turn out one way or another. We are swayed by our genes, not fated by them. Genetics, we can conclude, isn't destiny. It isn't even Destiny's Child.

We realise now that nature and nurture don't act in isolation. Musically gifted children, for example, are likely to have musically gifted parents who are likely to provide their offspring with both the genes and the upbringing to help musical talent flourish. But there's more to it than that. Our DNA, it seems, may help shape qualities such as musical aptitude, appreciation and motivation, which in turn may help shape the desire to practise and perform. This in turn may attract praise from a parent, a teacher or a screaming crowd, which in turn may help shape the desire to practise and perform … and so on. Nature and nurture don't go it alone. They interact with one another. The environment isn't something that just happens to us. Because of our genes, we may be more likely to search, avoid, flourish or flounder in certain situations. 'We select environments that are correlated with our genetic propensity,' says geneticist Robert Plomin from King's College London. 'People create their own experiences, in part, for genetic reasons.' So it wasn't his DNA that predetermined that life-changing moment in 1953 when Elvis walked into Sun Records to cut his first record, nor was it those long evenings singing on the front porch with his family. Instead, Elvis's genes *and* his environment had been dancing an intricate bossa nova for all of his life.

With identical DNA, any variants related to musical ability that Elvis had would also be in GElvis's genome. By immersing him in music, his parents could help GElvis to play to this innate genetic strength, but it would be no guarantee of rock 'n' roll stardom. From twin studies we must draw the conclusion not that our DNA dictates our talents, but that we should encourage our children to try different things, find out what they enjoy and what they are good at, and then support them all the way. With a favourable genetic undertow, they will then be more likely to excel. Finding your mojo is about learning to play to your genetic strengths, not flogging a dead genetic nag.

Identical But Different

Armed with an intimate knowledge of his own genome and the variants in it, GElvis might choose to use this unusual heads-up to his advantage. Knowing that he carried variants linked to disease and had a 'twin' with health problems, he might decide to do things differently. He might exercise regularly and choose low-fat spread instead of peanut butter. He might ditch the burgers and have a doctor keep an eye on his cholesterol. One thing's for certain, though; the world that GElvis would grow up in would be very different to 1930s Mississippi. And that would lead to another subtle layer of biological differences that in turn, could influence the way he turned out.

It's called epigenetics and trust me, it's a great word to throw into pub banter if you're looking to impress your friends. Epigenetic changes are those that affect the way genes work without ever altering the sequence of the DNA itself. So identical twins with identical DNA are not the same epigenetically. They may have the same genes but the way those genes are switched on and off is different.

The most commonly studied epigenetic change is one called methylation, where a tiny bundle of atoms called a methyl group attaches to any one of 20 million different locations in the genome and switches genes off. Inexpensive, off-the-shelf 'chips' can be used to reveal where the methyl groups land, and in the last few years there has been a surge of studies comparing the methylation patterns of identical twins. Of particular interest are twin pairs where one sibling has a particular characteristic, schizophrenia or cancer for example, and the other one doesn't, and what tends to be found is that the two siblings have different methylation profiles. It's jumping the gun to presume that these changes *trigger* schizophrenia or cancer or any of the other characteristics to which they have been linked. It is, of course, more complicated than that. For most of these disorders, we don't yet know if the methylation patterns are the cause or a consequence of the disease. But what we do know is that the environment is a big driver of epigenetic variability.

Pesticides, pollutants, alcohol, smoking and diet all influence our epigenetic profile, which in turn affects gene activity, sometimes with lasting effect. The experiences of pregnant mothers, we realise, can shape the epigenomes of their unborn offspring. In one well-studied strain of mouse, the Agouti yellow, subtle changes in the mother's diet determine whether her offspring will turn out thin with brown fur, or fat with yellow fur. And in another classic animal study, researchers found that the pups of neglectful, inattentive rat mothers ended up with more methylation around key stress-related genes, turning them into skittish, jumpy, stressed-out adults.

People who smoke have different methylation profiles to non-smokers, and although when they kick the habit their profiles return to 'almost' normal, certain changes can last for decades. It's thought these 'hard to shift' methylation patterns could help explain why ex-smokers remain at an increased risk of cancer and respiratory problems years after they've stubbed out their last cigarette. The experiences of a pregnant mother can influence the epigenome.

Life experiences shape methylation patterns, which in turn influence the activity of our genes, which in turn have the potential to shape our behaviour, lifestyle choices and health. Identical twins, some studies have suggested, start off with similar epigenetic profiles, which then become more different as the years go by and the siblings live increasingly separate lives. But what's amazing is just how different the epigenomes of identical twins can be at birth. Jeff Craig and Richard Saffery from the Murdoch Childrens Research Institute, Melbourne, have studied identical twins born prematurely and found that their epigenetic profiles are already different at 32 weeks – a full two months before most 'regular' babies are born. This is because although the unborn siblings may share the same environment – their mother's womb – the experiences they encounter inside it are different. One twin might be squashing the other or be closer to the mother's heartbeat. One twin might have a slightly narrower or longer umbilical cord, skewing the flow of nutrients from mother to child. Although the differences might seem insignificant, Craig believes they have the power to wreak significant change. Talk to the parents of identical twins and they'll very often tell you how their babies' personalities were different from day one. Very often identical twins are born different weights. Sometimes they are born with different hair or eye colours, or one twin might be healthy while the other has some sort of medical problem. Early life experiences, linked to epigenetic change, could be responsible for some of these changes. And the effects could be long-lasting, too. Craig suspects that sometimes, the time we spend in our mother's womb may lay the biological foundation for diseases that we develop decades later.

Transferred not as a twin but as a singleton into the womb of a surrogate mother, the first nine months of GElvis's life would be unavoidably different to that of his DNA donor. Instead of competing for resources and space like normal twins, GElvis would have his embryonic abode all to himself. And where Presley's mother lived in poverty in 1930s America, GElvis's surrogate would enjoy an altogether more contemporary lifestyle, with plentiful food and state of the art medical care, all of which would drive variation in the epigenome of her unborn baby. After birth, as GElvis grew up in this twenty-first-century world of ours, his diet, family, friends and experiences would be unlike those of his famous clone. This would drive further change in his epigenome, and although the exact repercussions are impossible to predict, it's safe to say that they would conspire to make Elvis and GElvis less alike, rather than more similar. But more than that, Craig and colleagues have found epigenetic differences in cells taken from naturally conceived twins and twins created via IVF. His results suggest that the very act of creating a child artificially in a lab may be enough to drive epigenetic change. A product of gene editing and cloning, GElvis would be conceived not under a duvet, but in a petri dish. Far from identical, Elvis and GElvis would be epigenetically different from their initial moment of creation.

An Infinite Number of Elvises

So, all of us – Elvis and GElvis included – are a product of nature and of nurture, with epigenetics bridging the gap between the two, but it's still not enough to explain how we turn out the way we do. There's another factor at work.

In 2013, Gerd Kempermann of the Dresden site of the German Centre for Neurodegenerative Diseases and colleagues took 40 genetically identical mice and raised them all in the same identical environment. If nature and nurture are all there is, he reasoned, then each of the mice should turn out the same. But it didn't quite pan out that way.

The team housed their mice in a five-storey rodent 'des res' kitted out with flower pots, plastic tubes and toys, then recorded their movements over a three-month period. What they found was that although the patterns of movement were similar at the start of the experiment, by the end of it, they were very different indeed. Some animals were bold and explored a lot, while others had no such wanderlust. 'The animals developed different personalities,' says Kempermann. Genetically identical pups grew into very different adults.

Kempermann thinks the differences are a consequence of the way each individual animal interacted with its environment. A slightly more active mouse, for example, might explore more than a less active one. It might bump into more of its cagemates, climb on a flower pot, tumble down a tube. This might fuel its adventurous spirit, make it better at climbing, more likely and able to seek out new experiences. All of the mice lived in the same environment but they experienced it very differently. 'Over time, the rich environment lost its sameness,' says Kempermann.

It's the same for children growing up in the same family. 'If you ask parents if they treat their kids differently, they say no. But if you ask their children the same question, you'd swear they're growing up in a different family,' says Robert Plomin. '"It's not fair!" is the norm.' Supposedly shared environments, such as home and school, and common life events, such as birth and bereavement, are experienced differently not just by twins, but by all family members. It's this non-shared environment, unique to the individual, that must have caused Gerd Kempermann's mice to end up different, and that helps to shape each and every one of us, too.

In addition, our lives are full of uncertainty and chance. Bill Clinton, famously, was inspired to go into politics after shaking the hand of John F. Kennedy. And I, less famously, was inspired to take up writing many years ago after I injured my knee dancing to Sinitta's 'So Macho' in a Glaswegian nightclub, and for months was unable to go out partying with my friends. Serendipity, unpredictable by its very nature, plays a major role in the way our life stories are written. Who knows what would have had happened if Presley's vocal cords had been nobbled by a random throat infection in his early years? Or what if Elvis's natural twin,

Jesse, had survived? Perhaps the boys would have been too busy scrapping or trading bubble gum cards to take much interest in music. The fact is, even if you created an infinite number of GElvises, turned back time and brought them up in a carbon copy of Presley's 1930s America, they'd still all end up different.

Jailhouse Rock

Although it's technically possible to decode Elvis Presley's genome from his quiff, Presleyfy a human cell and create a baby from it, that child will never 'be' Elvis. At this point, depending on where they work, the scientists involved could have broken several laws, not to mention social and ethical taboos. Under the UK Human Tissue Act of 2004, it is illegal to analyse, without consent, the DNA of someone that has been dead for less than 100 years. A somewhat arbitrary cut-off, this means that researchers at the UK's University of Leicester who are working to decode the genome of a different king – King Richard III – are several hundred years in the clear, while anyone trying to sequence Elvis's genome on British soil could find themselves doing the 'Jailhouse Rock'.

Human reproductive cloning is widely banned (see Chapter 2) and in 2002, after an in-depth review on the subject, the President's Council on Bioethics wrote to the then US president, George W. Bush, to advise him that 'cloning-to-produce-children is not only unsafe but also morally unacceptable, and ought not to be attempted.' It smacks of eugenics, relegates the creation of human life to a manufacturing process, and leads to a weird and uncomfortable world in which a father could become 'twin brother' to his own 'son'.

To top it all, the tiny, cloned human embryo would have been genetically modified. Every cell in the developing clone's body would be derived from the original Presleyfied egg, which means that when the clone grows up and thinks about having tiny Presleys of his own, the three million laboratory-engineered changes would also be present in his sperm. These changes would then be passed on from one generation to the next.

It wouldn't be the first time scientists have edited heritable genetic changes into human embryos. In April 2015, Chinese researchers announced that they had used CRISPR to repair a defective disease-causing gene in human embryos. The preliminary research, performed deliberately on non-viable human embryos (because the scientists never planned to let the embryos grow into babies) was a step towards trying to correct the single mutated gene that causes beta-thalassemia, a potentially fatal blood disorder. Despite the good intentions, the research played into a heated debate. A month earlier, researchers writing in the journal *Nature* called for a global moratorium on the genetic modification of human embryos, citing 'grave concerns' over ethics and safety concerns that it seems were justified. When the Chinese researchers ran their experiments, the gene editing didn't go as well as they had hoped. Sometimes CRISPR failed to cut the DNA, other times it cut it in the wrong spot, potentially creating new problems. The technique only worked properly in a fraction of the embryos tested and when it did, the embryos sometimes ended up being a mix of edited and non-edited cells, and not the pure genetically modified embryo of design. As present technology stands, genome editing, it seems, is not without risk.

So we create GElvis despite the grave societal and moral peril to us and him, and for what? To create someone that looks a bit like Elvis but who, inevitably, will be a different person with a different life story. Cells can be cloned. 'Selves' cannot. From the moment it is created, every new life is unique, and it only becomes more unique as time goes on. And that, my friends, is the wonder of you.

Elvis truly was one of a kind. Geneticist Neil Hall from the University of Liverpool has calculated how many more babies Gladys and Vernon Presley would have to have had before they produced one that was genetically identical to Elvis. His calculation takes into account the random divvying up of the two sets of 23 chromosomes that happens when cells divide to make sperm or egg, as well as the swapping around of genetic material within chromosomes that accompanies the process. He estimates that Gladys and Vernon Presley would have to produce another 41, 000,000 children before Gladys popped out a natural Elvis clone. That's 41 followed by 126 zeros. 'This is a ridiculously large number and quite possibly a massive underestimate,' says Hall. It also doesn't take into account the fact that after a third child, married couples rarely have sex again, or that Gladys and Vernon are, in fact, dead. To put the number in context, the number of atoms in the observable universe is thought to be around 4 with 79 zeros after it. 'Presley's number,' as it shall henceforth be known, is 48 zeros bigger than that. It's mind-bogglingly huge, and a testament to the uniqueness that was Elvis.

There is, however, another way to bring some much-needed 'Elvisity' back into the world. Let me explain. When Elvis died, in 1977, it's estimated there were around 170 Elvis impersonators. By the year 2000, that number had increased to around 85,000. Plot the figures on a graph, take into account projected population growth, do some dubious stats and what we find is that by 2043, 1 in 4 of us will be an Elvis impersonator. If you're reading this book on a bus please look at the person sitting next but two to you. Now imagine them with a glistening, three-inch quiff. Now stop staring. I think they've noticed. More excitingly still, extend the line on the graph to its logical conclusion and what we see is that by 2050, every living man, woman and child on the planet *will* be an Elvis impersonator. We may be able to 'de-extinct' a person with the same genetic code as Elvis, but it won't *be* Elvis. We can't recreate Elvis. We shouldn't recreate Elvis. We don't NEED to recreate Elvis. His music is still with us and one day soon we're all going to be singing it. Ladies and gentlemen, put on your jumpsuit, fasten your cape and fashion your hair into an Elvis-style quiff. You have a little over a decade to learn the lyrics to 'Suspicious Minds'. But in the meantime, Elvis really has left the building.

CHAPTER SEVEN

Blue Christmas

Elvis may well have vacated the premises, but as research progresses, the prospect of successfully resurrecting extinct species is still very much alive. One day, that 'almost' gastric-brooding frog tadpole will very probably turn into a frog. And if it's not the woolly mammoth or the passenger pigeon that makes a comeback, it will be some other extinct animal. With the birth of Celia's clone, the little bucardo kid who lived for just seven short minutes, we briefly de-extincted life, but as science progresses, boundaries will fall. It took researchers decades to amass the fundamental knowledge and technical know-how needed to make Dolly the sheep, but two decades later cloning has become an everyday occurrence. At Sooam Biotech in Seoul, South Korea, scientists produce 500 cloned animals every day, including dogs, pigs and cattle. Cloned animals are being produced as pets, as tools for medical research and for use in the livestock industry. It's not a big leap to imagine that the hurdles that cut the cloned bucardo's life short will be overcome. It's not a case of 'if' de-extinction will happen, but 'when'.

It's important, then, to think about the issues this dogma-challenging breakthrough will raise. De-extinction isn't about creating lonely zoo exhibits, it's about producing sustainable animal populations that will thrive in the wild. With that in mind, we would be wise to consider exactly what these animals will be like. Could they be dangerous? Will they help or hinder the environment they are released into? How would they be managed and protected? And how to choose what the best candidate or candidates might be?

Never Be the Same

So let's start by getting one thing straight. A de-extinct animal will *never* be the same as the original. These will be twenty-first-century creations, not the finely sculpted products of millions of years of evolution. Just like all of the Elvis impersonators in the world, they will be proxies, not replicas. Because of the methods used to create them, there will be fundamental differences in their biology. And because they will be born into a world that never stands still, their environment will have changed. A product of nature, nurture and the interaction between the two, all this will conspire to make any de-extinct animal unavoidably different to its extinct counterpart.

In South Africa, scientists are using back-breeding to de-extinct the quagga. A relative of the living plains zebra, the quagga had a stripy front half and a plain rear end. So they're choosing the zebras that look most like quaggas and letting them breed. The aim, over successive generations, is to create animals with drearier backsides, but will they be quaggas? Well, we've been selectively breeding dogs for hundreds of years and although the resultant cockerpoos,* collies and chihuahuas may look different, they are all still pooches. In a similar vein, a twenty-first-century quagga will be a new breed of zebra, not a new species.

When George Church is done editing mammoth genes into his elephant cells and has used them to create an animal, he will end up with a beast that may look like a mammoth but whose genome is still overwhelmingly elephant. It will be a genetic 'cut and shut', a well-intentioned mixture of DNA from two different species. Likewise, the genome of Ben Novak's passenger pigeon will be predominantly band-tailed pigeon with just a genetic 'hint' of the bird that he loves so much. Even the bucardo clone, whose genome was 99.9995 per cent pure, was still not 100 per cent bucardo.

DNA aside, other aspects of biology will be different, too. The environment, remember, leaves physical marks on DNA in the shape of tiny chemical groups that bind to it and alter gene activity – so-called epigenetic changes. Created artificially, raised in the twenty-first-century, the epigenome of a de-extinct animal will be different from the original. And there's another '-ome' to consider, too. All complex animals are at least equal parts bacteria, equal parts their own cells. This collection of microbes, known as the 'microbiome', will also be different … more on this later.

If a de-extinct creature turns out to be healthy, and looks the same and acts the same as the real McCoy, you could argue that these 'invisible' differences matter little. However, they are important as they are likely to influence everything from whether or not the animal survives in the first place, to how it will be treated when it is released into the wild.

---

*Cockerpoos, arguably the best type of dog on the planet, are a cross-breed between a cocker spaniel and a poodle. Their name, however, highlights the alarming trend towards linguistic blending that has seen the rise of words such as 'flip-flocks' (the unacceptable wearing of flip-flops and socks), 'cellfish' (a person who uses their phone in the quiet carriage of a train) and 'askhole' (someone who asks too many stupid questions).

Tears of a Clone

Genetic and epigenetic differences are thought to be a big part of the reason why so many cloned embryos fail to develop normally. European scientists had to make hundreds of bucardo clones in order to achieve just one live birth, which then died within minutes (see Chapter 3). And it's a similar story when researchers have tried to clone endangered species, like the banteng (*Bos javanicus*) and the gaur (*Bos gaurus*).* All of these projects involved the same method – interspecies cloning – where the DNA of an endangered or extinct species is reprogrammed by the egg of a closely related, living species. The resulting embryo has the nuclear DNA of the cloned species but the mitochondrial DNA of the donor egg, and it's this fundamental mismatch, researchers suspect, that causes so many of the clones to fail.

Molecular biologist Rhiannon Lloyd from the University of Portsmouth suspects the problem may lie in a tiny structure found on the inside wall of the mitochondria. Known as the mitochondrial oxidation phosphorylation system, it's a complex of interacting proteins that work together as a molecular machine to generate energy for the cell. 'It's incredibly intricate,' says Lloyd. But the component protein parts are encoded by genes from nuclear *and* mitochondrial DNA. So if the nuclear and mitochondrial DNA come from different species, the proteins won't necessarily mesh together properly, and the machine won't work. 'Even small perturbations can lead to problems,' she says. Without energy, cells can't survive. And if the cells run out of steam, then a developing embryo will fail.

But there's another difficulty. A big one. And that is that no one really understands how cloning works. When life first starts, the cells inside an embryo can turn into any of

---

*The gaur is not, as I had originally thought, an Area of Outstanding Natural Beauty in South Wales but in fact a very large, very muscly type of Asian ox. Bantengs are another species of Asian cattle.

the many different cell types in the adult body, but as the embryo develops and specialised cells are formed, this ability becomes lost. The different types of cells – brain cells, heart cells, skin cells and the like – are all still genetically identical, but different genes have been switched on and off. Their epigenomes are different.

So when cloners transfer a nucleus into an egg, they transfer not just the nuclear DNA, but the epigenetic instructions that control it. Something inside the egg, and no one's exactly sure what, then has to silence these instructions; remove the signals that say 'be a brain cell' and instead give it the liberty to become any cell type it chooses. It's like restoring the factory settings on your mobile phone. Sometimes this feat is achieved and the clone develops normally. But other times the DNA isn't reprogrammed fully and that, researchers think, is one of the reasons why so many cloned animals fail to survive.

Gut Feeling

When scientists studied a baby mammoth called Lyuba that was discovered in the Siberian permafrost in 2007, they found bits of dung in her stomach. Baby elephants, we know, eat their mother's droppings because it helps them establish the gut bacteria they need to digest their food. Mammoths, it seems, did the same thing. It would have contributed to the animals' unique blend of bacteria – its microbiome.

When we come into the world, our microbes are few, but as we pass through the birth canal and begin to sup our mother's milk, our own characteristic blend of bacteria starts to be established. It's fine-tuned further by what we eat, where we live and what we do. Trillions of bacteria live in and on our bodies, and such is their importance it's led some to claim that our collective microbial community be recognised as an organ in its own right. Studies have revealed that the microbiome plays a major role in both health and disease. Changes in the human microbiome, for example, have been implicated in many disorders including diarrhoea, diabetes and depression. So when it comes to de-extincting an animal, perhaps we should be thinking about de-extincting their unique blend of microbes, too.

Depending on what is left of the extinct creature, this will be more or less difficult. When scientists examined the remains of 'that bloody mammoth', the gnarled beast with seemingly liquid blood and fresh, frozen tissue discovered in Siberia in 2012, they found it still had faeces in its rectum. Microbiologist Bas Wintermans from the VU University Medical Center in Amsterdam had the enviable job of retrieving them and then trying to determine their microbial content. He's found that the dung contained a mix of regular, twenty-first-century intestinal bacteria and some as-yet unidentified, possibly Ice Age microbes. It raises the possibility of being able to inoculate a de-extincted mammoth with the gut bacteria of its ancient forebears. 'When the time comes for the mammoth to be de-extincted, then we will design a suitable intestinal microbiota for it,' says Wintermans, 'but I think that this will take a while.'

Microbes, of course, come in different varieties. While a healthy blend of gut bacteria is good for us, disease-causing bacteria and viruses are less welcome. There is anxiety that by de-extincting an animal, we might accidentally de-extinct some of the tiny pathogens that once infected it, which could pose a danger to public health. It's a valid concern. In 2014, European scientists revived a virus that had lain dormant in the Siberian permafrost for over 30,000 years just by thawing it out and giving it something to infect. Fortuitously, the virus, dubbed *Pithovirus sibericum*, turned out only to infect amoebae, but it raised the possibility that other, potentially more infectious viruses could be lurking in the frozen wastelands of the North. The next year they found another, different virus in the same permafrost sample. As the world warms and the

Arctic thaws, we are likely to find more of these pathogens, but it's not necessarily a cause for alarm. Just as the DNA of Ice Age animals will have broken down over the thousands of years they have remained frozen, so too will the genetic material of any viruses or bacteria that once blighted them. You'd expect an Ice Age microbe to be as dead as the Ice Age animal it once used to infect. It could be the case that the recently discovered Siberian viruses were unusually resilient or there was something particular about the way they were frozen that helped their genetic material survive. Either way, it's possible to recognise genetic sequences from viruses and tell them apart from the DNA of other species. When scientists initiate the first step in de-extinction, decoding the genome of the animal of interest, they will be able to screen for any suspicious viruses, then check again when they have made their embryos, then re-check when they have their animals. Just as strict precautions should be in place to safeguard the health and welfare of the animals, so too cautionary measures should be in place to ensure that the animals are not carriers of disease. De-extinct animals would begin their days in a state of captive quarantine where their health and any potential problems could be monitored. They would only be released into the wild when the relevant regulatory agencies had deemed it safe to do so, and if regulatory agencies are known for anything, it's for bureaucratically taking their time.

What's in a Name?

Suppose, however, that scientists pull it off. They manage to de-extinct a species and set it free. Next comes the problem of how to classify it. If the animals we create through de-extinction are unavoidably different to the originals, then what should we call them, and how should they be categorised? It might, at first glance, seem little more than an exercise in semantics, but the labels we give these animals will have profound repercussions on their future wellbeing.

It's certainly a conundrum for the International Union for the Conservation of Nature (IUCN), compilers of the Red List and the organisation currently producing guidelines on how to classify de-extinct animals. Species on the Red List find themselves grouped in different categories of hope or hopelessness, ranging from 'least concern' and 'vulnerable' all the way to 'endangered' and 'extinct'. To qualify for legal protection under endangered species law, an organism must first be listed as endangered, but here's the rub ... the Red List criteria only apply to animals living in the wild. Created and raised in captivity, the first de-extinct animals would not initially be wild, and thus might find themselves excluded from the Red List. It's a bizarre situation. Initially one of a kind until more of the same are engineered, the existence of this new type of animal would suggest a certain precariousness. Yet it's only when the animal is released into the wild that it can qualify for Red List status and protection under the Endangered Species Act.

It could, however, qualify as something else. If the animal has been created through cloning and/or had its genome edited, it would probably be classed as a 'Genetically Modified Organism' (GMO), but the related legislation varies widely around the globe and tends to focus more on GM foods, medicines and pesticides than it does GM animals. This makes things a little uncertain. Suppose scientists bring back the woolly mammoth and decide to set it free. In Europe, robust risk assessments would have to be made before the GMO was released. In England, researchers would have to apply for a licence from Natural England, the governmental advisory group tasked with protecting the country's nature. If granted, this would probably come with restrictions. If the animal wandered outside of its licensed range, for example, there could be instructions to have it destroyed. Someone had better warn the mammoths.

Maybe they should apply for a Green Card, as in the United States GMO legislation is much more permissive. There is no relevant over-arching federal law. Instead there are bits of laws, mostly drawn up for other purposes, which vary around the country. Three Californian counties, for example, have regulations banning GMOs from their jurisdiction, but in the vast majority of the country, there is nothing to stop someone from creating and releasing a herd of woolly mammoths into their own backyard. 'It's quite conceivable you could release a de-extinct animal in your own property and have it spread into the wild – and in the US at least, you're not liable under any national government law,' says bio-law expert Andrew Torrance, from the University of Kansas.

In another unexpected twist, the creators of a de-extinct animal might find themselves in a position where they don't actually own the creature into which they have invested so much time and money. In the United Kingdom, and in parts of the United States, the 'law of capture' may apply. If a woolly mammoth were to wander into my back garden and I put a lead around its neck, it might officially be mine … which would please the children, but might put a strain on neighbourly relations. And things could be strange for de-extinct animals in captivity, too. When China exports its pandas to zoos in other countries, the recipients sign a contract agreeing that China still owns the animal. Were a similar situation to occur with de-extinction, Pleistocene Park might one day find itself 'renting' woolly mammoths from the United States or South Korea or Japan, rather than owning them.

As to what literally to call the animals we de-extinct, it's all up for grabs. All the living things that we know about have both a common name and a scientific name. One suggestion mooted by the IUCN is that we give a de-extinct animal the scientific name of the surrogate species, followed by the name of the original form and then some indication of the method used to produce it. So George Church's mammoth would become '*Elephas maximus Mammuthus primigenius CRISPR-Cas9 Interspecies Somatic Nuclear Transfer*' and Novak's passenger pigeon would be '*Patagioenas fasciata Ectopistes migratorius CRISPR-Cas9 Primordial Germ Cell Transfer*'. It's all a bit of a mouthful. Common names, in contrast, are easier on the tongue and can vary between people and languages. It's here then, that we can have some fun. I propose Church's hairy, snow-loving elephant be called a 'woollyphant', and that Novak's feathery friend be called 'Passenger Pigeon II: The *Ectopistes* Strikes Back'.

Decision Time

So if you could de-extinct one animal, who or what would you choose? Let's recap the options discussed so far. Dinosaurs and dodos are out because there's no DNA to be had. The thylacine project has folded for lack of interest (shame!). The bucardo project has stalled for lack of funds. Woolly mammoths are on the cards but their creation by necessity will involve experimentation and invasive procedures on its endangered living relative, the Asian elephant. Neanderthals (and Elvis Presley) are off limits because their de-extinction would be ethically wrong, pose a risk to human health and is ultimately pointless. With chytrid running riot, there's currently nowhere wild for the gastric-brooding frog to go. And there's reason to suspect that in order to survive, the passenger pigeon would have to exist in flocks so big it might become a public nuisance.

When people are trying to decide which living species to conserve, they often plump for the iconic, beautiful and charismatic ones. Think pandas, tigers and gorillas. They take a great photograph, but they're not necessarily the ones most deserving of our attention. For every celeb species, there are millions of disenfranchised wannabes that never make the front page. Who cares for the Titicaca water frog (*Telmatobius culeus*), a South American amphibian whose ill-fitting, saggy folds of skin have earned it the nickname of the 'scrotum frog'? Who among us has heard of the pineapple sea cucumber (*Thelenota ananas*), the chainsaw-nosed narrowsnout sawfish (*Pristis zijsron*) or the Yoda-faced pied tamarin (*Saguinus bicolor*)? All are in dire straits, yet for whatever reason fail to make the homepage of the World Wildlife Fund or filter into the public consciousness.

It's a similar story with de-extinction. Scientists tinkering with the technology tend to focus their efforts on species that are memorable, eye-catching and iconic. These might be the ones that are physically big, like the mammoth, or that we have the deepest emotional connection with, like the passenger pigeon. The bucardo, so recently extinct, is part of Spain's cultural heritage, while the gastric-brooding frog, currently the focus of a de-extinction project in Australia, is so unusual that it jumps out at you. These are all valid reasons for choosing 'this' or 'that' species, but if we're seriously to pick one candidate for de-extinction, there are other factors to consider.

We need to be practical. Our choices are limited by what is technically possible. First, we need a plentiful source of DNA, but because the molecule degrades over time that instantly restricts the list of possible candidates from those that lived in the last 3.5 billion years (when life first evolved) to those that were alive in the last million years. Next, we need a close living relative to act as a reference genome. Its genetic sequence can then be used to help order the ancient DNA fragments and plug any gaps. So although DNA has been retrieved from the moa (*Dinornis giganteus*), a huge flightless bird that went extinct around 600 years ago, its closest living relative, a much smaller bird called the tinamou, is just too different to provide a useful genetic template.

Later in the process, when the ancient genome has been assembled and coaxed in to life inside a cell, the living relative will need to step in again. If it's a marsupial, it must carry the embryo in its pouch, if it's a placental mammal, it must let it grow in its womb, and if it's a bird it must lay its egg. Here again there are restrictions. The closest living relative of Steller's sea cow, for example, is another marine mammal called the dugong. But a newborn Steller's sea cow would be the same length (about 2–3m; 6–10ft) as an *adult* dugong. It'd be like asking a dachshund to give birth to a Great Dane.

In some cases, the same surrogate would then have to nurture the de-extinct animal not just through embryonic development, but after birth, too. So we need to consider if the surrogate would make a decent parent. As well as being able to feed the young animal, the parent might also need to teach the youngster practical skills. But would an elephant be able to teach a woolly mammoth how to pluck grass from under the snow, or show it where the ancestral migration paths are? In some cases, humans might need to step in.

Consider the California condor (*Gymnogyps californianus*). This huge, bald, beaked vulture is a scavenger, but 30 years ago it was driven close to extinction by lead poisoning from the bullets lacing the carrion it ate. In 1987, conservationists decided to round up all the remaining wild birds and use them to establish a captive breeding colony. To maximise the number of youngsters produced, they removed and hand-reared the first egg laid by a pair, causing the female to lay a second egg, which the parents then raised on their own. It was a hugely labour intensive but successful endeavour, and thanks to their work there are now more than 200 California condors soaring on the thermals above California, Arizona and Mexico. But they still need our help. Although a partial ban on lead ammunition was introduced, condors still suffer from lead poisoning so they're captured from time to time, and, if needed, treated with a drug to purge the lead from their bloodstream. At any one time, around 10 per cent of the wild population is in veterinary respite. Another problem stems from their impressive three-metre (10-ft) wingspan. Conservationists realised to their horror that the homegrown birds were getting electrocuted when they accidentally flew into power lines. So now the birds are also given cable-aversion therapy. Artificial poles placed in large aviaries train the birds to avoid cables by giving them a non-lethal but memorable electric shock.

When animals are returned to the wild after a period of absence, we can't presume that we will just let them go and all will be fine. We need to think carefully about the problems they might encounter and the level of help they will need. De-extinction will be a long process. It won't stop with the birth of a bucardo kid or the hatching of a passenger pigeon. It will continue through the captive breeding phase, into their wild release and beyond. The de-extinction process isn't just about 'making' the de-extinct animal, it's about ensuring there is a place for it and giving it the best start so that one day it will be able to thrive on its own.

All De-extinct But Nowhere to Go

For some animals, the lack of suitable habitat makes their de-extinction a complete no-go. The Yangtze river dolphin (*Lipotes vexillifer*), for example, once graced the freshwaters of the Yangtze and neighbouring Qiantang River in China, where legend has it that the balletic swimmer was the reincarnation of a drowned princess. But then China started doing what China does best: making things. The Yangtze River became one of the world's busiest, most polluted waterways, and the river dolphins ended up colliding with boats, getting tangled in fishing nets and drowning. As of 2006, when a scientific survey failed to find them, they've been presumed extinct. The degradation of their natural habitat means there's no point bringing them back. The banks of the Yangtze are lined with sprawling cities and an estimated 12 billion cubic metres

(423 billion cubic feet) of untreated waste is pumped into its murky waters every year. This river is too dirty for a princess.

If we're going to choose an animal to de-extinct, it'd better be one that has somewhere to go, and it would be helpful if we knew exactly what drove the species to extinction in the first place. That way we can be sure the threats have either gone or can be mitigated. Look closer into the story of the Yangtze river dolphin and the causes of its extinction become as murky and convoluted as the waters it swam in. The river was dirty and polluted and full of boats and nets, but upstream the then newly built Three Gorges Dam also prevented the fish that the River Dolphin ate from migrating to their spawning grounds. It's a complicated picture with multiple confounding issues. 'We don't know which one or ones of these factors was the cause of the Yangtze river dolphin's extinction,' says Samuel Turvey from the Zoological Society of London, who studied the creature. In the meantime, the Yangtze is becoming more busy, polluted and fished. 'The whole system is a terrible mess,' he laments.

The Wonderful World of Christmas

The Christmas Island rat (*Rattus macleari*) is a promising candidate for de-extinction. It was suggested to me by ancient DNA researcher Tom Gilbert from the University of Copenhagen, who has been involved in sequencing its DNA, and it makes for an interesting case study. This large brown rodent was once endemic to the island after which it is named, a biodiverse dot in the Indian Ocean, 240 miles south of Java. But in 1899 the SS *Hindustan* weighed anchor and black rats, hiding in the hold, jumped ship and scampered onto dry land. Shortly after that, the native Christmas Island rats were seen staggering around as if drunk, and four years later they had all disappeared. But they hadn't gone into 'rehab'. Black rats, or rather the disease-carrying micro-organisms called trypanosomes that lurked in their fleas, were blamed, but no one was able to prove this until Gilbert and co. took a look at the creature's DNA. They extracted DNA from museum specimens and found that Christmas Island rats alive before the black rats' invasion were trypanosome-free, but animals that died post-contact were very often infected.

Here, then, is an animal for whom we know the cause of extinction; an invasive species inadvertently introduced by humans. Its DNA is accessible from the many museum specimens that exist, and it has a close living relative in the form of the humble lab rat, whose biology and genome are well studied. Rats have been widely used as experimental animals since the early nineteenth century because they are small, amenable animals that reproduce quickly. They were also among the first mammals to be cloned, all of which provides a useful pool of species-specific knowledge that will help anyone trying to bring the Christmas Island rat back. Here is a creature whose de-extinction is technically possible and whose demise we inadvertently caused. So perhaps we do have a moral obligation to bring it back. It's no poster child, so would be flying the flag for those forgotten underdog species. And the fact that rats would never win first place in a popularity contest might even work in the researchers' favour. Cloning and de-extinction are wasteful processes, but who cares if another rat embryo bites the dust? The Christmas Island rat could be used as proof of principle, to see the de-extinction process through from its inception to its ultimate endpoint, the ongoing survival of a de-extinct species in the wild. When sufficient numbers had been bred in captivity, the Christmas Island rats could be re-introduced to their original home, where around one-third of the island's 50 square miles is protected as a National Park. Surrounded by water, their range would be naturally restricted. They couldn't swim to Java and wreak havoc with its natural wildlife. And if the rats ever did become a pest on their home turf, we could always cull them. We've ridded numerous islands of problem rats before. It might not be easy, but it is possible.

It sounds like the perfect choice. But here's the problem. Ecosystems don't stand still. They are dynamic systems in a constant state of flux. Superficially the rainforest of Christmas Island may look the same as it did 150 years ago, but it's not. Other species have come and gone.

Although the island is still home to a plethora of unique and fascinating native wildlife – the stunning annual migration of the Christmas Island red crab (*Gecarcoidea natalis*) happens here – the island has been battered by wave after wave of invasive species. Every 5–10 years, supercolonies of non-native yellow crazy ants (*Anoplolepis gracilipes*), so called because of their erratic movements, build up. Left unchecked, they swarm over the native red crabs and spray formic acid into their eyes and mouthparts. In the last few years, they've killed a quarter of the island's crabs, an act that has not gone unnoticed by the rest of the ecosystem. Left to their own devices, the red crabs dig burrows, turn over the soil and fertilise it with their droppings. They eat leaf litter, seeds and fruit. But as their numbers have dwindled, more seedlings are sprouting. Weeds are creeping into the forest and as the flora changes, so too do the types of animal that flourish or flounder in it. Put simply, the flora and fauna of Christmas Island have changed.

Despite being the perfect candidate in so many ways, it's unclear how the Christmas Island rat would fit into this scene. The trypanosome-infested black rats that drove it to extinction are still there, as are a plethora of other potentially troublesome non-native species. The Christmas Island rat could find itself staggering around, ravaged by disease, having acid squirted in its face. It could go extinct all over again.

Perhaps, then, it could be put it on a different island, a well-established conservation principle called 'translocation'. Islands are water-fringed dots of biodiversity. They harbour a fifth of all species, including some that are totally unique.

Invasive rodents, however, can be found on over 150,000 islands where they are one of the most widespread and damaging of alien species. If rats are known for anything it's for scaring girly women, and being resourceful and resilient. They are scavengers, opportunists and disease-carriers that all too easily displace other species. As a result, introduced rodents have been responsible for approximately half of all known extinctions in the last 500 years. The de-extincted Christmas Island rat might not just 'survive', it might do so well it becomes an invasive species itself.

If de-extinction is to complement conservation and boost biodiversity, then we'd better think carefully about the repercussions of our actions. There's no point going to all the trouble of de-extincting a species if we later regret our actions and have to destroy it. If our long, embittered battle with invasive species teaches us anything, it's that single species can have a huge effect on their environment. All living things are part of an ecosystem, a biological community of interacting organisms and the physical environment they live in. Change one component of that ecosystem and it's like playing Giant Jenga with a toddler. If you're lucky, the tower wobbles a bit and stays standing, but if you're not the whole thing can come crashing down. Living things are important, not just in their own right but because of the roles they play within their ecosystems. Some are scavengers, some are grazers, others carnivores. There are pollinators, seed dispersers, water purifiers and pest controllers. Then there are the damn builders, the burrowers and the decomposers. The list goes on. All living things do ecologically relevant 'jobs' that help maintain the ecosystem they live in. If we're going to choose an animal to de-extinct, it'd better be one whose ecology is well understood. Extinct within a decade or so of human settlement, scientists never really got the chance to study the Christmas Island rat. We know it was a scavenger and suspect that it helped keep the red crab population in check. But we don't know for sure, and in the time it's been gone, the roles it once performed have more than likely been taken by the rash of invasive species that currently plague the island. What started as a promising front-runner turns out, on closer inspection, to be a non-starter.

Wolf Call

In the last few decades, conservationists have broadened their focus from trying to save individual species, to trying to preserve ecosystems. When a species disappears, we realise now that it leaves more than a physical gap. It leaves a job vacancy. Vital ecological roles become lost, causing ripples that can be felt elsewhere. If an animal or plant low down the food chain is wiped out, then the species that depend on it for their own survival start to feel the pinch. Similarly, if an animal at the top of the chain disappears, there are trickle-down effects all the way to the bottom. It's called a 'trophic cascade' and over time these changes can re-sculpt landscapes.

Ninety years ago the wolf (*Canis lupus*) went extinct, not globally but locally. Its job, as a top predator, made it unpopular with locals and in 1907 the US Fish and Wildlife Service sanctioned an extensive pest control programme that saw the animal exterminated from Yellowstone National Park. But its disappearance did not go unfelt. In the years that followed, the ecosystem changed radically. Without wolves to compete with them, coyote numbers increased, which in turn led to a decrease in the numbers of their favourite food, the pronghorn antelope. Without wolves to predate them, elk populations rocketed and the species that they ate – deciduous woody trees like aspen, willow and cottonwood – became grazed to within an inch of their lives. That meant there were fewer places for birds to nest and less wood for beavers to build their dams with. So their numbers in turn declined. With fewer beaver dams, the water table dropped, making it even harder for trees to grow. And with fewer tree roots to stabilise them, the riverbanks and hillsides became increasingly eroded. Yellowstone changed. A lot.

The wolf proved to be a keystone species, an animal that has a disproportionately large influence on the environment relative to its abundance. Like a keystone in a bridge, when the species is removed the whole structure crumbles. Then in 1995, after a very long and very heated debate, the wolves came back. Fourteen of them, captured in the Rocky Mountains of Western Alberta, Canada, were transported to Yellowstone, where they were later released. What happened next exceeded expectation. Wolves didn't just influence the number of elk in the park; they changed the animals' behaviour. The elk learned to stay away from the valleys and the ravines where the wolves could trap them, and in time these areas regenerated. Woody plants regrew. Trees shot up. Their roots stabilised the riverbanks so they collapsed less often, causing the rivers to become more fixed in their course. There were more songbirds. The beavers came back, built dams and created ponds that provided habitat for fish, amphibians and reptiles. The wolves killed coyotes and as a result, the numbers of small mammals increased, which attracted hawks, weasels and foxes.

Wolves, the much-maligned bad boys of children's fairy tales, 'did good'.

Of course, they weren't *that* good. They were still wolves, after all. From time to time they strayed into neighbouring fields and ate the odd sheep, making them unpopular with farmers (and sheep). But when it came to boosting biodiversity, they were a howling success. The story of the wolf demonstrates how sometimes the re-introduction of a single species can have a very large and very positive impact. It is an ecosystem engineer. Its presence helps to build, sculpt and maintain ecosystems, creating opportunities and physical niches where other species can thrive.

The passenger pigeon was also an ecosystem engineer. When the vast flocks came to rest, they decimated the forest. Boughs were broken and trees were toppled. The closed canopy woodland was transformed into an open, guano-covered wasteland. But this apparent devastation created life. The droppings fertilised the soil, and sunlight was able to reach the new grasses, flowers and shrubs that grew. This provided more habitat and resources for insects, reptiles, birds and mammals. The trees that remained sprouted new shoots. The landscape became productive, diverse and bioabundant, until the closed canopy regrew and the cycle started again. Passenger pigeons drove this continual recycling and rejuvenation of the eastern North American forests. In their absence, regeneration has virtually stopped. The forests are stagnant and populations of native animals are in decline. Bring back the passenger pigeon, the argument goes, and the natural cycles of biodiversity-boosting forest regeneration could be brought back, too.

So perhaps rather than thinking about which species we should de-extinct, we should ponder instead what job vacancies there are. What ecological 'holes' are there that need plugging? What roles are left unfilled? We shouldn't de-extinct an animal just because we can or because we miss it or because we feel morally responsible. There needs to be a deeper sense of purpose. De-extinction, judiciously employed, offers a means to fill these vacancies, to 'repair' ailing ecosystems so that biological interactions and services that have been lost can resume. It's about ecological enrichment, returning communities of animals to the wild where they can live and interact with one another, and positively influence their surroundings.

It's also worth adding one final thought here. When beavers were returned to the River Otter in Devon, UK, after a 400-year local extinction, like the Yellowstone wolves they too had a positive effect on local biodiversity. But beavers aren't just ecosystem engineers, they're also civil engineers. The 13 dams they built along one particular stretch have turned a stream that used to hold a few hundred litres of water into a resource that can contain 65,000 litres (14,500 gallons). So now when there's a sudden downpour, the water doesn't gush down the stream and flood the surrounding landscape. It's stored in the pools behind the dams then slowly trickles down the natural staircase the beavers have created. In a country where the current government is spending billions on flood management, beavers *could* be part of the solution. Arguments rage over whether the River Otter beavers should stay or go, but the idea is at least worth exploring. In the meantime, visitors come, hoping to catch a glimpse of these winsome wild animals. Local hotels and eateries are benefiting. Devon does a particularly nice cream tea, after all. There is potential for ecotourism. The judicious re-wilding of our natural world with animals, de-extinct or otherwise, could yield practical and economic benefits.

If I were to write a newspaper advertisement to help find the perfect de-extinction candidate it would read as follows:

WANTED: Extinct Animal with Lust for Life

Calling all stuffed, pickled, dog-eared and moth-eaten museum exhibits.

Are you fed up of being gawped at by sticky-fingered schoolchildren all day long? Is your body falling to bits but your DNA in great shape? Do you ever wish you weren't dead? Did *we* kill you? If you answered 'yes' to all these questions, read on …

Bring Back the King is on the hunt for animals with existing relatives and habitat to bring back from the dead. After de-extinction you will be treated to a short, all-expenses-paid stay in a captive environment of *our* choice. You will then be returned to the wild, where you will work for the benefit of the ecosystem and help turn the world into a more beautiful, biodiverse place.

Keystone species preferred. Ice Age animals considered. Dinosaurs need not apply, nor any other species that eats humans.

Disclaimer: The de-extinct version of you will not be identical to the pre-extinct version of you. Gene expression can go up as well as down. Always read the small print.

CHAPTER EIGHT

I Just Can't Help Believing

On 2 July 1982, US truck driver Lawrence Richard Walters tied 45 helium-filled weather balloons to a patio chair he had tethered to the ground and sat down. His plan was to cut the ropes and float gently above his Californian backyard at a height of about 9m (40ft) for a couple of hours. He'd have a beer, eat a sandwich and enjoy the view. Then when he'd had enough, he'd use his pellet gun to pop the balloons and bring himself back down to Earth. What could possibly go wrong?

But when his friends severed the ropes that anchored the Heath Robinson contraption, the chair didn't rise slowly. It streaked into the sky like a rocket, quickly reaching a height of 4,500 metres (15,000 feet) and drifted into the primary approach corridor of Long Beach Airport. In desperation,

Walters began shooting the balloons, but accidentally dropped his gun overboard. In the end, he had to wait for the chair to descend of its own accord, but as it did, cables dangling beneath it became snarled in an electrical power line. It caused a 20-minute blackout in the Long Beach area.

Walters, or 'Lawnchair Larry' as he became known, was able to climb down to the ground where members of the LAPD duly arrested him. As he was handcuffed and led away, waiting reporters asked him why he had done it, to which he replied casually, 'A man can't just sit around.'

Putting the soaring stupidity of his story to one side, Lawnchair Larry's adventure highlights an interesting dichotomy. Just sitting idly by isn't necessarily a beneficial or salutary course of action. However, on the flip side, doing something just because you can isn't necessarily a good idea either.

In *Jurassic Park*, mathematician Ian Malcolm points out that the dinosaur-creating scientists were so busy thinking about whether or not they *could* bring animals back from extinction, that they didn't pause to think whether or not they *should*. These are remarkably prescient words for a man whose body is so tightly clad in leather you would imagine it would be hard for him to breathe, much less utter pearls of ethical wisdom. But you have to admit he had a point. Just because we have the technology to edit genomes, create life in a dish and have one species give birth to another, doesn't necessarily mean it's a good idea. We are on the verge of bringing extinct species back from the past, but is it something we should be doing?

A Most Unusual Sperm Bank

When frog expert Mike Mahony received the phone call asking him to join the Lazarus Project he jumped at the chance, not just because he wanted to de-extinct the gastric-brooding frog, but because he realised there would be another very positive spinoff. At the time, all sorts of frogs in his native Australia were being driven to extinction by the chytrid fungus (see Chapter 5). Species were disappearing from the most pristine, protected habitats. One summer, he'd visit a particular stream and it would be full of frogs. The next year he'd go back and nothing would leap out at him. The stream was deserted. 'We were literally watching frogs go extinct and there was nothing we could do about it,' he says.

In desperation, Mahony had been devising a backdoor rescue: he had been catching endangered frogs in the wild, then collecting and freezing their sperm. The idea was that one day, the samples could be thawed and used for IVF to help make new frogs. Mahony's sperm bank was an insurance policy against future frog extinctions. But the fly in the ointment is that frog sperm are only half the deal. Frog eggs are also needed, but they are too large to freeze. They shatter. So Mahony's plan, by necessity, would involve marrying the thawed sperm of one species with the freshly laid eggs of a different, living species.

'That's where it got interesting,' he says, because the Lazarus Project also involves melding cells from different frog species, albeit via cloning. The two projects had obvious parallels. Knowledge gained from the Lazarus Project could, Mahony realised, help him preserve species that were still clinging to life. Through their efforts to bring back the gastric-brooding frog, researchers are accruing vital knowledge about frog reproductive biology. They've learned how to stimulate ovulation and how to physically handle and manipulate the eggs that are laid. They've learned about tissue culture conditions, and how to give freshly created embryos the best chance of survival. From de-extincting the gastric-brooding frog to conserving living frog species, these are all transferable skills. 'The Lazarus Project has helped us develop technologies that help to prevent extinction,' says Mahony. De-extinction isn't just about bringing back the dead; it's also about helping the living.

De-extinction has the potential to help conservation in two ways. In its early stages – where we are now – it's contributing indirectly. It's producing knowledge that is useful for those trying to save endangered species. This is what Mahony has found. In its later stages – when de-extinct animals are released into the wild – it will have a direct effect. The de-extincted species will make for greater biodiversity, and if an animal proves to be a keystone species then the effects will be even more profound. 'Conservation,' Wikipedia tells me, 'is an ethic of resource use, allocation and protection. Its primary focus is upon maintaining the health of the natural world, its fisheries, habitats and biological diversity.' De-extinction could be one of those resources.

Conservationists, as the moniker might imply, tend to be a pretty conservative bunch. There are many among their ranks who don't like the idea of de-extinction. They argue that it will steal attention and funding from attempts to conserve the living. If we can bring species back from the dead, perhaps it will become easier to let existing species slip away. The incentive to prevent extinction might become devalued, and 'sitting around' spectating a species' demise tacitly condoned. What will it matter if the mountain gorilla (*Gorilla beringei beringei*) or the leatherback turtle (*Dermochelys coriacea*) disappear if we can bring them back again later when we feel like it?

These are valid concerns, but for now that's all they are – concerns. The field of de-extinction is literally and metaphorically in its embryonic phase. We don't know how or if de-extinction will change our feelings towards the natural world, but it could work as a positive force. Imagine scientists make a mammoth and reveal it to the world for the first time. It could be a 'man on the moon' moment, witnessed by billions, remembered by generations to come. Rather than making us care less about wildlife, de-extinction could make us care *more*. It could inspire people to become scientists, conservationists and protectors of wild places. We may have woolly mammoth fossils and cave paintings, we may even have their carcasses frozen in the Arctic, but there'd be nothing quite like meeting a living one. The same goes for any extinct species brought back to life.

Think about the passenger pigeon. We may have colourful prints in American field guides, but these 'book-birds' would pale into insignificance next to the riotous beating of a billion pairs of wings as they darkened the sky and dimmed the daylight.

No one is suggesting that we stop protecting our natural world or the creatures that live in it. This is not an 'either or' scenario. De-extinction is something to be developed and then used alongside other conservation methods, not replace them. Nor is de-extinction siphoning funds away from conservation. Heavyweight conservation hitters such as the World Wildlife Fund aren't putting any money into any de-extinction research at all. At the time of writing, work on the bucardo has stalled for lack of moolah, and the passenger pigeon project continues on a wing and a prayer with funds of an uncertain future. At the Sooam Biotech Research Foundation in South Korea, attempts to de-extinct the woolly mammoth are only made possible through the organisation's more-profitable endeavours, such as dog cloning, while in the United States, George Church's attempts to 'mammothify' the elephant genome are only possible because the gene-editing technology being used has another, much more fundable purpose – to help find cures for human disease. De-extinction isn't holding conservation upside-down and forcibly shaking the money from its pockets; it's struggling to make ends meet.

I've interviewed a lot of conservationists while writing this book, and the impression I get is that they're nervous of high-tech fixes. 'Wildlife groups and NGOs don't want the answer to be in a test tube,' says Mike Mahony, 'because they don't want their central message diluted.' But technology is already having a positive influence on conservation. A broad range of techniques already exist. At one end of the spectrum, methods such as vaccines and genetic tests are widely accepted and increasingly used. Santa Catalina's diminutive island foxes (*Urocyon littoralis*) were saved from the canine distemper virus that was decimating their numbers after they were systematically captured and inoculated with a specially developed vaccine. In Nepal, simple DNA-testing of tiger scats offers a non-invasive way of determining not just the identity and gender of an individual animal, but how genetically diverse – and potentially how resilient – the tiger population is. Researchers can then use this information to make practical conservation decisions, like whether or not to move animals between different populations. In the United States, they've gone a step further. Scientists there have sequenced the full genomes of 36 California condors and now use this more detailed genetic map to direct their captive breeding programme (see Chapter 7). It's the first species for which this has ever been done. In the past, some condors were born with a genetically determined form of lethal dwarfism, but the genomic registry means that carriers of the disease can now be identified and excluded from breeding.

Methods such as these are no-brainers, but at the other end of the spectrum, a variety of more invasive techniques involving assisted reproduction are being explored. In the bird world, for example, primordial germ cell therapy is being developed (see Chapter 4). Sperm from the threatened houbara bustard (*Chlamydotis undulata*) have been grown inside a chicken and then used to generate live houbara chicks, raising hopes the technique could be used to boost the numbers of other endangered birds, too.

In other parts of the animal kingdom, cloning is being explored. In 2015, for example, Mohammad Nasr-Esfahani and his team from Iran's Royan Institute successfully cloned an endangered sheep known as the Esfahan mouflon (*Ovis orientalis isphahanica*), which is found only in a very small area of east-central Iran. DNA from a mouflon cell was inserted into a regular sheep egg that had had its own nucleus removed, and the resulting embryo was then transferred to the womb of a surrogate sheep. The result was a healthy, happy lamb named 'Maral' (which means 'beautiful') who, when I spoke to her creators, was still thriving many months after her birth.

This second batch of techniques is more intrusive, expensive and experimental than the first, all of which makes them divisive. They are also, as you may have spotted, exactly the same methods as those being used for de-extinction. The intention is to revive the passenger pigeon through primordial germ cell technology, and the woolly mammoth and the gastric-brooding frog are to be brought back via interspecies cloning.

Although it's not initially obvious, there is, I think, a blurry line between conservation and de-extinction. They're not discrete entities. They overlap at the extreme end of the conservation spectrum, where the same high-tech methods may be able to boost the numbers of species alive and dead. José Folch and his team cryopreserved cells from Celia, the world's last bucardo, just 10 months before the falling tree killed her. Suppose for a moment that they had started their cloning experiments earlier and success had been more immediate. If that first little clone had been born before, rather than after Celia died, it probably would have been called an act of conservation, another example of an endangered animal being produced through interspecies cloning. But the end result – a cloned bucardo kid – would have been no different. But it could also just as easily qualify as an act of de-extinction. From a functional perspective it didn't matter that Celia was still alive; the bucardo went extinct the moment it became unable to breed its way out of trouble. De-extinction, 'extreme conservation', call it what you will, is all about trying to enhance biodiversity, and history shows us how sometimes radical measures are called for. Consider the tale of the black-footed ferret.

Audrey Hepburn with Whiskers

The black-footed ferret is a dainty, photogenic North American mustelid that looks like Audrey Hepburn in animal form. It has kohl-like smudges around its eyes, a perfect button nose and, as its name suggests, looks like it has trotted through a puddle of ink. It's not extinct, but there was a time when people thought it was. No stranger to controversy, the little carnivore has been involved in a conservation story that spans decades. It's the perfect example of how innovative solutions can sometimes change the fate of a species.

In the 1950s, people thought the species was no more. It seemed to have vanished from its native habitat, the Great Plains of North America. But then, in 1981, near Meeteetse, Wyoming, a dog called Shep made an unexpected discovery. He brought his owners, John and Lucille Hogg, a gift of a dead black-footed ferret. After much discussion, it led the US Fish and Wildlife Service, the organisation responsible for overseeing American wildlife, to make an unprecedented decision. They resolved to go out and capture every single last black-footed ferret they could find and bring them all into captivity where they could be kept safe and allowed to breed. 'It was a historic decision,' says conservationist Samantha Wisely from the University of Florida, 'no one had ever done this before.'* It was a move that didn't sit well with traditional conservationists, who felt strongly that wild animals belonged in the wild, whatever their predicament. Nevertheless, 11 males and seven females were caught and began new lives in captivity.

The scientists knew they would have to manage the breeding programme carefully in order to make the most of this limited genetic diversity. Members of the same species may share the vast majority of their DNA, but subtle differences in their genomes – genetic variation – are what endow populations with the ability to weather change. When a population shrinks or is descended from a limited number of founders, genetic variation becomes eroded; descendants become more inbred, making the future of any

*They did the same thing a little later with the California condors.

species less certain. In a world that's warming, where habitats change and diseases come and go, species need this genetic chutzpah to see them through. So they adopted a strategy of 'arranged marriages', carefully orchestrated matings between select individuals designed to keep the offspring as genetically vibrant as possible. But despite their best efforts, certain animals still ended up being massively under-represented in the gene pool. For reasons unknown, some males just weren't interested. So a lady who earned herself the unenviable nickname of 'The Sperm Queen', the late great veterinarian JoGayle Howard from the Smithsonian Conservation Biology Institute, decided to try artificial insemination. It was another bold intervention because the technique used to place the sperm close to the females' wombs was new, and involved laparoscopic surgery. Thanks to Howard's hard work, the risk paid off and more than 140 black-footed ferret kits have been born through artificial insemination. Bolstered by a carefully managed natural breeding programme, over 9,000 captive ferrets have been born in total over the last 30 years, of which around half have been released back into the wild.

But the species isn't out of the woods yet. There are still two major problems that need to be overcome. First, not all of the original founding members were able to reproduce (not even artificially). That means the entire black-footed ferret population alive today is descended from just seven of the 18 original members. 'Everybody is a second cousin,' says Wisely, who has studied their genetics. And second, the disease that caused their demise in the first place, the sylvatic plague, is still out there. If captive management and artificial insemination were seen as gutsy choices, then what's being proposed next is even more audacious.

When Ryan Phelan and Stewart Brand founded Revive and Restore in 2012 they were frustrated with the way things were going in the world of conservation. 'I've always been very fond of wildlife,' Phelan tells me one sunny May evening as we chat by phone. 'The problem is that conservation has a limited set of tools in its box. For a long time, it's been about protecting land and the species in it. It's not been about using state-of-the-art genetic technology. I wanted to bring genomics to conservation.' So Revive and Restore isn't just about de-extinction. It aims to enhance biodiversity by applying genetic methods to extinct *and* endangered species. Through Revive and Restore, Phelan and Brand are nurturing the science needed to make it happen, and are dedicated to fostering public support. It was they who organised the 2013 TEDx de-extinction event. Now, working with dozens of scientists, they are participating in five de-extinction projects (including Church's Woolly Mammoth Revival and the Great Passenger Pigeon Comeback), and six that aim to prevent the future extinction of endangered species. The black-footed ferret is one of these.

In collaboration with San Diego Zoo Global and sequencing company Cofactor Genomics, Revive and Restore has sequenced the genomes of four black-footed ferrets. Two of the samples came from living animals, but the other two came from the cryopreserved tissues of ferrets that were captured with the original founders but that died without leaving living descendants. Preliminary results suggest that the living animals may indeed be suffering the effects of inbreeding, and that the two dead animals contain unique genetic variants not found in the living population. If cells from these animals were used to make cloned ferrets it would return these lost variants back into the gene pool, effectively increasing the size of the founding population from seven to nine. It could be enough to make all the difference.

It's a proposal that will make some people start to twitch. Critics of 'conservation' cloning point out that the procedure yields more failures than successes, and can't possibly produce animals in anything like the quantity needed to assuage the current biodiversity crisis. Regular ferrets, however, were first cloned back in 2006, and are commonly used in medical research. We have a good grasp of their reproductive biology, so the chances of success are good.

When it comes to conservation and de-extinction, cloning isn't about churning out huge numbers of genetically identical animals. There'd be no point. They'd end up more inbred than sandwich filling. It's more subtle than that. We live in a time where genome sequencing has become affordable and accessible. Cryopreserved cells and museum specimens are libraries of genetic variation that can be read like a book. A recent study of DNA from living tigers and museum specimens found a huge number of genetic variants that were present in historical specimens but that are absent in the depleted, living population. If researchers can catalogue this variation from museum specimens and other sources, there's nothing, in theory, to stop them from editing it into any animals they create. We don't need to create an entire business of genetically identical ferrets, when an extra couple of genetically eclectic individuals would do.* These could inject some much-needed variation back into the black-footed ferret gene pool, but there's no point doing it if the animals are then set free and die from sylvatic plague.

So alongside San Diego Zoo Global, Revive and Restore are submitting two proposals to US Fish and Wildlife. The first recommends boosting the black-footed ferret's gene pool through cloning, and the second suggests the animal have its genome altered to make it plague-resistant. 'If these two projects work,' writes Stewart Brand in the 2015 Revive and Restore Year End Report, 'they will make conservation history. Neither has been attempted before, and what is learned from the black-footed ferrets could be applied to countless other species in need of genetic rescue.'

---

*A 'business' is the correct collective noun for a group of ferrets. Other favourites of mine include a gulp of cormorants, a flamboyance of flamingos and a knob of waterfowl.

Genetic Rescue

These pioneering strategies seem like a logical extension to an existing conservation practice. For years, animals have been deliberately moved from one place to another so that they can breed and introduce new genes into existing populations. It's the original, low-tech way to jazz up a population's genome. By 1986, the world's population of the Norfolk Island boobook (*Ninox novaeseelandiae undulata*), a reddish brown owl that once thrived on a couple of islands near New Caledonia in the Pacific Ocean had dwindled to a single female. But then conservationists imported two male New Zealand boobooks (*Ninox novaeseelandiae*). Love blossomed. Eggs hatched and the rest, as they say, was history. Similarly, the Florida panther (*Puma concolor coryi*), once depleted in numbers and seriously inbred, was rescued after eight cats of a related subspecies, the Texas panther (*Puma concolor stanleyana*), were introduced to its range and the animals interbred.

'We've been doing this for a long time,' says conservation biologist Gary Roemer from New Mexico State University, 'but then we began to wonder, instead of moving the animals, why don't we just move their genes instead?' Roemer and colleagues fleshed out the idea in a 2013 thought piece written for the journal *Nature* entitled 'Tweaking Genes for Conservation'. They call the process 'facilitated adaptation', because they'd be using genetic techniques to help animals adapt to change. Revive and Restore, which was independently considering the same idea round about the same time, calls it 'genetic rescue'. Both groups are thinking about how the genomes of different species could be deliberately and precisely altered to improve their chances of survival. Just as we move towards an era of 'precision medicine', where treatments are tailored for individuals on the basis of their individual genetic make-up, so too we could move towards a time of 'precision conservation' ... if that's what we decide to do.

In the plant kingdom, such genetic tinkering is well practised. Today, 12 per cent of arable land in the world is planted with genetically modified (GM) crops and the GM seed market is valued at US$15 billion. In the field of conservation, the American chestnut tree (*Castanea dentata*) has had its genome modified to make it resistant to an invasive fungus that was driving the species towards extinction. William Powell and Charles Maynard from the State University of New York cut and pasted a fungus-resistant gene from wheat into the genome of the American giant, creating a blight-hardy plant that can pass its resistance on to future generations through its seeds. The tree has proved its worth in test sites, so the next step is to gain regulatory approval and 'release' the tree into the wild. So if it can be done for plants, why not animals?

What if gene-editing technology could help save the Tasmanian devil (*Sarcophilus harrisii*), a misrepresented cartoon caricature of an animal that is currently being wiped out on its home soil by a contagious cancer that is spread when the animals bite each other? Perhaps bats could have their DNA tweaked to make them more resistant to white-nose syndrome, a deadly fungal disease that infects the mammals as they hibernate and has killed more than five million bats in North America. In Hawaii, native birds are being trounced by a form of avian malaria that is transmitted by a single species of mosquito, *Culex quinquefasciatus*. In tandem with habitat loss, it's turning the island into an extinction hotspot. People have tried and failed to rid the islands of the pest using insecticides, so perhaps it's time to start considering other options. Vaccines are tricky, because the parasite is excellent at hiding from the host's immune system and there is currently no effective way to administer vaccines to wild birds. The answer may lie with a trial being run by British biotech firm, Oxitec. It is looking into eradicating the mosquitoes that transmit human dengue fever with a method that involves altering the genes of the males, then releasing them in large numbers into the wild. The idea is that they will breed with normal females and that their offspring will inherit an engineered gene that prevents them from developing into adults. If something similar could be done for the mosquito that carries avian malaria, it could perhaps offer salvation for the unique birds of Hawaii, or at least help slow the disease's spread.

Similarly, de-extinct species could be engineered with a genetic heads-up. Not all frogs, for example, are killed by the chytrid fungus. Some manage to beat the infection. Researchers have studied the DNA of these animals and found they carry variants in a group of genes involved in the immune response. If these presumed 'resistance genes' were cut and pasted into the genome of a gastric-brooding frog during the de-extinction process, when the 'almost tadpole' turns into a frog, it might stand a fighting chance back in the wild.

How about we think bigger? Global climate is changing. The world is warming. A few years ago, German researchers identified versions of genes that help the commercial rainbow trout (*Oncorhynchus mykiss*) to survive in warmer waters. These gene variants could be inserted into the genomes of fish eggs or embryos in populations threatened by rising water temperatures. It could help them to survive. If George Church is altering the elephant genome to create an animal that is better adapted to the cold, what's to stop us from tweaking other genomes to make them amenable to changing temperatures?

Genomic technology has the potential to profoundly influence the future of the world's wildlife. From a technical perspective, it's not the genetic modification *per se* that's the difficult bit. Complex characteristics, like being resistant to a particular disease or being able to survive temperature change, will be controlled by many different, interacting genes. 'Trying to identify the genes that are responsible will be a major hurdle,' says Roemer. This will take time.

Playing God?

Another hugely important element will be getting the public to buy in to the idea. The proposed genetic fixes are new and unfamiliar. De-extinction is untried and unknown. Not everyone is in favour. The scientists involved in these new techniques have been accused of 'playing God', the products of their work dubbed 'unnatural'. Environmental ethicist Ben Minteer from Arizona State University believes that de-extinction in particular represents a refusal to accept our moral and technological limits in the natural world. 'We have this impulse to constantly tinker and manipulate nature,' he says. 'It can be pathological. I've thought about it a lot, and it crosses a line. I worry about the sense of inevitability of all this.'

It's important to address these concerns. Before 1978, IVF was in the same boat. It was new and unfamiliar. Many people thought it was unnatural and would create monsters. Patrick Steptoe and Robert Edwards, the British researchers who pioneered the technique, were accused of 'playing God'. But then Louise Brown was born, the world's first test-tube baby, and attitudes began to soften. Today Louise has a son of her own, and around the world, more than five million test-tube babies have been born. IVF is no longer seen as 'freakish' or 'unnatural'. It's a widely accepted fertility treatment that has brought immeasurable joy to millions of childless couples.

Similarly, before the mid-nineties mammalian cloning was an unknown. Ian Wilmut and co., the team behind Dolly, were accused of 'playing God', and some worried it was the start of a slippery slope towards human reproductive cloning. Then Dolly was born, and attitudes began to change. The agricultural industry gradually adopted the technique, while human reproductive cloning became widely banned.

Sometimes methods that are initially perceived as new, artificial and unnerving become accepted as they become familiar and are proven to be safe and helpful. As for the claim that these techniques are unnatural, Ben Novak, lead scientist on the Great Passenger Pigeon Comeback, points out that all the techniques developed by us for assisted reproduction and genome editing stem from naturally occurring counterparts. Parthenogenesis, a form of cloning, is known to occur in some shark and snake species, not to mention turkeys and Komodo dragons. Sperm storage, where females delay fertilisation by stowing semen in their reproductive tracts after sex, is relatively commonplace in fish, reptiles, birds and amphibians. For some bat species it means that a couple can mate in the autumn but postpone the birth of their young until spring. It's the natural equivalent of artificial insemination, not so different to human couples that decide to have semen stored for later use. The precision gene-editing tool of the moment, CRISPR-Cas9, was not invented by geneticists but adapted from a primitive bacterial immune system. And we're not the first to move genes between species. Hybridisation, where one species breeds with another, occurs frequently in the animal kingdom. 'While we have certainly designed our own innovation for reproductive and genome technologies, we are not truly gods,' says Novak.

Scientists are not making these suggestions on a whim; they're proposing them because there are wildlife problems that are not being solved by the methods we have. Along the way, the research that is being done is generating knowledge that may be of value to us all. It's about understanding how life begins and how embryos develop, how a single cell can somehow morph into a fully formed animal. By understanding the processes that guide embryonic development, we strive to gain some sort of influence over them. So when things go wrong, we can help. Research into the earliest stages of life is helping to drive the development of new treatments and therapies for humans. The Iranian team behind the cloned mouflon, for example, are using their experiences of cloning different mammals to generate cloned goats that make drugs such as insulin in their milk. And the man who worked out how to have one bird produce the sex cells of another – Robert Etches, now at Crystal Biosciences in California – is modifying the technique to produce chickens that make therapeutically useful antibodies in their eggs.

Of course, just because we can do something doesn't mean we automatically should; Lawnchair Larry showed us that. His homemade helium-powered flying machine was, with hindsight, a dreadful idea. But the rather casual rationale for his aerial exploits illustrates an innate truth. 'A man can't [or shouldn't] just sit around.' This course can be dangerous.

Every year, around US$8 billion is spent conserving biodiversity, yet despite our best intentions the number of endangered species continues to grow. We have to accept that sometimes traditional conservation methods aren't enough. The pioneering spirit that saw the world's black-footed ferrets rounded up and given assistance to breed in captivity is to be applauded. It almost certainly saved the species from extinction, but it was a short-term fix. We need new tools. De-extinction and genetic rescue could be part of the solution, but if we don't develop the science needed to make them happen, we'll never be able to make a genuine assessment of their worth. It's up to us to help decide what happens next. There's clearly a line that needs to be drawn between what is acceptable and what is not. It's vital that scientists engage with the wider world and respect and respond to the concerns that people may have. Their work needs to be transparent, carried out to the highest, most scrupulous of ethical standards. The public must be kept in the loop and have a chance to influence things. In the meantime, genetically modified animals – de-extinct or otherwise – are not about to flood our wild spaces any time soon. There's a lengthy and stringent process of ethical approval that will have to be gone through first, not to mention a lot of basic science that needs to be done. But surely it's worth at least considering their possible role in our rapidly changing world.

No one wants the answer to the world's biodiversity crisis to be found in a sterile laboratory. We want the answer to be in the natural world; in a fertile rainforest, a scorching African plain, a deciduous woodland or a garden pond. We want our wildlife to reproduce naturally, the way it's been done for millions of years, the way it evolved to happen. But in some cases it needs our assistance. I think that we should be keeping our options open, not dismissing any potentially helpful technology just because it's unfamiliar or 'artificial'. If we keep doing what we've always done then we're going to lose a lot more species than if we get off our lawnchairs and start exploring other options.

CHAPTER NINE

Now You See It ...

As flights go, it has to be the most memorable I have ever made. I checked in at Heathrow, waved goodbye to my suitcase and was hugely relieved when the stewardess noticed my heavily pregnant belly and thoughtfully bumped me up to first class. With ankles as swollen as mine, there was, after all, little chance of slotting them into an economy-sized footwell. At the time, I was working as a reporter for the journal *Nature*, and was travelling to Berlin to attend the 2004 European Society of Human Reproduction and Embryology conference. My remit was to find out what was going on in the world of assisted reproduction, stem cells and the like, and then write it up as news stories for the magazine's website. But 7 miles (39,000 feet) up, I started to become acutely aware of the person sitting next to me. He had seemed normal enough when I squeezed past him to sit down – middle-aged, well-dressed, clean-shaven – but now he sat silently staring straight ahead, clutching a robust silver briefcase to his chest.

'Can I put that in the overhead locker for you?' asked a passing stewardess.

'No thank you,' replied the gentleman, tightening his grip. 'I like to keep it with me.'

What, I wondered, could be so precious that he couldn't be parted from it? What cargo, hidden inside the hand luggage, could be so important or fragile he cradled it like a mother holding a newborn? Was it some priceless family heirloom, a delicate antique vase, or could it be something more sinister … a handgun or maybe a bomb?

The clinking of the drinks trolley disturbed me from my thoughts. I accepted a small miniature bottle of vodka, then noting that my neighbour had clocked my obviously expectant frame, began to babble. I had no plans to drink the alcohol, I told him. I had only accepted it because it was free and you have to accept free stuff on aeroplanes. That's why there's a cupboard in my living room that contains 87 sick bags. We started chatting. He was, he explained in a gentle Irish lilt, a medic and a researcher also on his way to the same meeting, where he was hoping to attract attention to his work.

'What work is that?' I asked nonchalantly.

He smiled and glanced at his briefcase. 'I can show you if you like,' he answered without a hint of mischief. Then he laid the case flat on his lap and carefully flipped the latches. As he lifted the lid I could see that the inside was padded with a thick layer of black foam, and there in the middle, in a specially moulded depression, was what I can only describe as 'an object'.

'The object' was roughly the same length and thickness as a family-sized bottle of fizzy drink. It was silver and smooth, fashioned from some sort of bright, shiny metal. Rounded at one end, it had wires dangling from the other.

He picked it up carefully and turned the hefty curio round in his fingers. Then he leaned towards me and placed it in my hands. 'Do you have any idea what this is?' he asked.

I'd like to think that, at that moment I came up with either the right answer or a witty, memorable reply. In hindsight, I can tell you that 'the object' looked like a piece of modern art or an enormous bullet. It could have been a dibber for an over-sized sapling or a huge metal cigar for an enormous metal robot. But I'm not very good on the spot. Instead, I just looked dumbfounded and said 'Err ... dunno.'

What he said next has to be one of the most remarkable sentences ever uttered by anyone in the whole of the history of life on Earth.

A Piece of Cake

Round about the same time, 200 miles (350 km) away, a young rhino called Fatu was celebrating her fourth birthday. At the Dvur Kralove Zoo in the Czech Republic, the birthday girl was chowing down on a specially prepared cake made of watermelon, apples, carrots and grass. Almost full-grown, weighing as much as a family-sized car, her hair-fringed ears twitched back and forth with delight as she tucked into her birthday treat. Next to her, her mother, Najin, and her aunt, Nabire, selflessly ensured not a morsel went to waste. And then, with the cake finally demolished, the rhinos went back to doing what rhinos do best: kicking back in the mud.

It was a time of bittersweet optimism. In the wild, Fatu's kind had been all but exterminated, while in captivity, less than a dozen remained. Hopes were high that in a couple of years, when Fatu reached sexual maturity, she would breed with one of the few remaining males and help boost the numbers of one of the most endangered species on Earth: the northern white rhino (*Ceratotherium simum cottoni*). Little did the birthday girl know, but heavy expectations had already been placed on her broad and muscular shoulders.

The northern white rhino is a magnificent beast. Built like a Russian tank, its thick, leathery hide hangs in folds over its large and powerful frame. It's not, as its name suggests, white, but rather a very subtle and beautiful shade of grey. It's the sort of colour paint manufacturers would call 'Shrew Whiskers' or 'Obelisk Grey' then charge you an arm and a leg to buy one small, watery tin. The word 'white' is thought to be a mistranslation of the Afrikaan's word '*weit*', which means 'wide', and refers to the animal's square-shaped muzzle. Dark eyes twinkle on either side of its long, wizened face and it has two curved horns on the end of its snout.

When it was first discovered, in 1907, the northern white rhino was common in parts of East and Central Africa. It roamed the savannas of Chad, Sudan, Uganda, the Central African Republic and what is now the Democratic Republic of Congo (DRC), where it used its uniquely shaped muzzle to mow the grass. But then people took a shine to its rather impressive horns and started to hunt it for so-called 'sport'. Commercial poaching set in. Little by little, relentlessly, the animal was exterminated from most of its range. But worse was to come. In the fifties and sixties, civil war broke out in Sudan and the DRC, and decades of lawlessness followed. In a world turned upside down by violence, it was impossible to keep the animals safe. The rhinos were slaughtered for their meat and for their horns, which in turn were exchanged for cash and weapons. First they disappeared from the Central African Republic, then from Sudan. In 2003, the year before Fatu's fourth birthday, just 15 northern white rhinos were estimated to remain in their final stronghold, the Garamba National Park in the DRC. A plan was hatched to transport some of the animals to safety in Kenya, but the local media branded the act 'a theft of national heritage' and politicians blocked the move. A few years later all of the wild northern white rhinos were dead.

Today, rhinos everywhere are in danger. There are five different species. Asia has the greater one-horned,* the Javan and the Sumatran rhino, while Africa has the black rhino and the white rhino, which can be subdivided further into northern and southern varieties. Three of the five different species are now critically endangered, and a staggering 95 per cent of the world's entire rhino population has been lost in just the past 40 years.

Dvur Kralove Zoo started importing and breeding different types of rhino in the seventies when it was clear that their future was precarious, and since then has found itself heavily involved in rhino conservation. In 1975, they imported six northern whites from Sudan. 'People realised that if the rhinos stayed where they were, it was just a matter of time before they would be killed by poachers,' says Jan Stejskal, Director of Communication and International Projects at Dvur Kralove. So Fatu's father, grandfather and four females made the 2,500 mile journey from the scorching banks of the Upper Nile to the more temperate Czech Republic. Two more animals were imported from elsewhere, and a little while later the not-so-tiny pitter-patter of rhino feet could be heard. Over the next two decades, four healthy northern white rhino calves were born, including Fatu's mother, Najin, and later Fatu herself.

But some rhino species, it seems, are harder to breed in captivity than others. The zoo, for example, has produced more than 40 baby black rhinos, but after Fatu, the northern white rhino calves stopped coming. No one knows why. The keepers tried everything: hormone therapy, a change in diet, different lighting. They even built a new animal house. Southern white rhinos were introduced into the northern whites' enclosure in the hope the two varieties might at least interbreed, but it was all to no avail. If the northern whites were unable to reproduce naturally then

---

*Which has the marvellous scientific name *Rhinoceros unicornis*.

perhaps, the experts at Dvur Kralove reasoned, something else would have to be done.

The 'Object'

Back on the aeroplane, I turned the weighty 'object' over in my hands. It was silver, shiny and smooth, and I found myself stroking it with my fingers, wondering what the hell it was.

'Go on then,' I said, 'put me out of my misery. Tell me what it is.'

But nothing could have prepared me for what the man next to me said. It was a combination of words I never expected could go together: a sentence, I am sure, I will never hear again.

'Well,' he said without a hint of the bombshell he was about to drop, 'I insert this up the rectum of rhinoceroses to make them ejaculate.'

Have you ever had one of those moments where time stands still? Where you are literally lost for words?

At this point in the story, I can tell you that one of three things happened. Can you guess which? Did I:

1. Shriek, 'What? You're telling me this thing's been up a rhino's arse?' so loudly that the passengers in adjacent rows asked to move seats?
2. Drop said 'object' on the floor from whence it rolled noisily all the way from first class down to the back of second, taking out an elderly lady who was on her way to the toilet?
3. Say 'how interesting', then politely hand the object back and surreptitiously sanitise my hands with the complementary vodka I was unable to drink?

Answer: I'm British. Of course, it was option 3. My neighbour, it turned out, was Dr Stephen Seager from the National Rehabilitation Hospital in Washington, DC, a man who has spent a large part of his career inserting probes up the rectums of not just rhinos, but of other animals, too.

Men who have suffered spinal cord trauma, or who have other illnesses, can sometimes struggle to achieve an erection and ejaculate. This can rob them of the ability to have children, but electro-ejaculation, to give it its proper term, can help. A probe of the appropriate dimensions is guided up the rectum and an electric current is then used to stimulate the nerves around the prostate gland, triggering ejaculation. The sperm can then be used for artificial insemination or *in vitro* fertilisation, or it can be frozen away and used at a later date. When I finally tracked Seager down, 12 years after our original encounter, he had only the faintest recollection of the incident. After all, he told me kindly, he does a lot of flying and frequently travels with his electro-ejaculator, but he was keen to talk to me about one of his proudest achievements.

Teenage cancer is a terrible thing. 'Six years ago,' he tells me, '95 per cent of boys in this age group diagnosed with cancer would have died. Now 95 per cent survive, but the chemo- and radio-therapy leave them sterile for life.' So Seager has adapted the electro-ejaculation technique so that it can be used on boys of this age. He collects semen samples from teenagers while they are being operated on for their cancer and, in so doing, offers them the chance to start a family later in life. Since he first started practising electro-ejaculation over 15 years ago, more than 100,000 babies have been born as a result of this technology. But it doesn't stop there.

Modified versions of these probes are now commonly used in animal husbandry. Remember that turkey you ate on Thanksgiving or Christmas Day? Chances are it was conceived via electro-ejaculation and artificial insemination. In the world of conservation, the technique has been used to produce offspring in a number of at-risk species including Przewalski's horse (*Equus ferus przewalskii*), the white-naped crane (*Grus vipio*) and the magellanic penguin (*Spheniscus magellanicus*). But there are two species for which the procedure has had a particularly profound effect. Most of the giant pandas (*Ailuropoda melanoleuca*) born in captivity have been produced via electro-ejaculation and artificial insemination, and the black-footed ferret's recent recovery from near extinction is in no small part due to the same technique (see Chapter 8). In his time, Seager has collected semen samples from leopards, tigers and polar bears, but when I met him in 2004, he was *en route* not just to the conference, but to Berlin Zoo to discuss how assisted reproduction might help their captive breeding programmes. 'I was worried that if I'd checked the briefcase into the hold, it might have got lost,' he tells me. 'Nowadays, of course, with all the extra security, I have to check my equipment into the hold. If I tried to take it on as hand luggage, I'd probably get arrested straight away. And then if I told them what the equipment was for they'd probably look at me and arrest me even more quickly.'

In the years that followed, Seager diversified and became more involved in the human side of things, while Berlin-based scientists have established themselves as world leaders in the field of large-mammal assisted reproduction. Veterinarian Thomas Hildebrandt and his team from the Leibniz Institute for Zoo and Wildlife Research (IZW) have spent the best part of two decades developing, refining and testing the equipment and procedures needed to help various animals, including rhinos and elephants, to reproduce artificially. A driven researcher and passionate defender of wildlife, he approaches his work with a scientific mind, tweaking a never-ending list of variables to improve the chances that the animals in his care will have offspring. Where possible he has patented his inventions, and he spends his days flying around the globe, visiting zoos and nature reserves, trying to change the fate of some of the world's most endangered species. So when Dvur Kralove's northern white rhinos stopped having babies, it was Hildebrandt that they called.

A Long Time Coming

I was intrigued to find out how the 'object' I encountered on my fateful British Airways flight is actually put to use. How exactly do you go about collecting semen samples from a two-tonne male rhino? So I too gave Hildebrandt a call.

Thomas Hildebrandt is a busy man, but very generous with his time. He speaks with a strong German accent, and proves to be candid, friendly and matter of fact. He explains to me how the probe he uses today is quite different from the one I would have seen on the aeroplane. 'It's a new concept,' he says. In the early days, the probe wasn't very reliable and sometimes caused damage to the animal, so Hildebrandt and his team reinvented it. The new version, based on their cumulative experience and encyclopaedic knowledge of the animals' internal anatomy, uses a lower voltage and is a slightly different shape. It's an altogether kinder, more efficient affair. It causes the animals no harm and it works every time.

Before it is used, however, the team must anaesthetise the animal and perform an ultrasound examination of its internal organs. It's a procedure they pioneered and have performed on more than 1,000 rhinos, so they know exactly what they are doing. Wearing surgical scrubs and shoulder-length gloves, Hildebrandt carefully guides a long, bendy laparoscope deep into the animal's rectum, then uses the ultrasound probe at its tip to look around. The black and white image is relayed to a nearby laptop screen, and thanks to a modified software program, reveals fine levels of anatomical detail. 'We know all the normal and abnormal structures that can be found in all five rhino species,' he says, 'and we can check for these.' The whole process takes less than 10 minutes, and only if the rhino is deemed fit and healthy do the team progress to the next stage.

After flushing out any urine and faeces from the rhino's bladder and rectum, it's time for the electro-ejaculator. The robust, cylindrical probe has been deliberately designed without a handle. It means that Hildebrandt has to plunge both the probe and the full length of his arm deep where the sun doesn't shine, but it helps him to make sure that the instrument ends up in exactly the right spot, near the top end of the urethra, close to the prostate. Then he nods to a colleague who flicks a switch and delivers a few short pulses of electricity to the device. 'Each pulse is no more than 15 volts,' he tells me, 'you can barely feel it on your tongue.' It's an image that makes me wrinkle my nose. With his arm and the 'object' still buried deep inside the rhino, Hildebrandt can't feel the pulses directly but he can sense the muscular contractions that they trigger. The animal begins to ejaculate but the intrusive intervention isn't over yet.

In the wild, males need a lot of stimulation to get things moving. A male rhino's penis is flanked by a pair of wing-like flaps that unfold inside the female and help keep the duo locked in a tryst, which can last up to two hours before the male ejaculates up to six times. Electrical stimulation brings the animal close ... but no cigar. Meaning that to finish the job, one of Hildebrandt's team must take matters into their own hands, literally. A colleague is called to 'assist' the animal, deliberately and vigorously massaging the penis so that when the animal does ejaculate the freshly released fluid can be guided along the pipework and into a collection tube. Each dose of ejaculate is around an espresso cup's worth of semen, but it contains millions of sperm; millions of chances for new life to be created. From start to finish, the entire ordeal takes no longer than an hour and when it is done, the rhino is roused from its anaesthesia and left to get on with the rest of its day.

It's a procedure that Hildebrandt performs around 40 times a year on different animals, including elephants, tigers and pandas. At the turn of the millennium, shortly after Fatu was born, he visited Dvur Kralove and collected and froze semen samples from two of the resident males. Then in 2006 and 2007, he returned and used the sperm to inseminate both Fatu and her mother, Najin.

Around this time, there was a flurry of good news. The first rhino ever to be conceived through semen collection and artificial insemination was born, not in the Czech Republic but in Hungary, at the Budapest Zoo. Then a year later another little rhino was born, this time produced with sperm that had been collected then frozen for three years before use. Here was proof that the artificial insemination procedures, so carefully honed by Hildebrandt and his team, really did work. And although both the newborns were southern rather than northern white rhinos, their arrival raised hopes that the technique would work in other rhinos, too.

Back in the Czech Republic, scientists waited patiently to see if either Fatu or Najin had become pregnant, but with the passing of time it became clear the inseminations had failed. The disappointing truth is that in most species tested so far, semen collection and artificial insemination fails more often than it succeeds. 'There is an impressive list of mammals for which artificial insemination has worked at some point,' says reproductive biologist William Holt from the Zoological Society of London, 'but the number of species for which this is now a routine method is extremely small.' Sometimes in rhinos, the technique begins to work – the inseminated sperm fertilises an egg creating an embryo, which then begins to divide. But then, for reasons unknown, division stops and the embryo becomes re-absorbed by the mother. What is known with certainty is that the longer female rhinos go without calving, the more difficult it becomes for them to get pregnant. It's a frustrating Catch 22. Without regular pregnancies, northern white females develop cysts in their wombs, which then make it even harder for them to conceive. That makes the fact that Hildebrandt has managed to create a total of seven baby rhinos through artificial insemination an impressive feat. 'It's an incredible achievement,' says the man from the plane, Stephen Seager. 'It's not easy to get semen from a rhino, it's not easy to inseminate a rhino and it's not easy to end up with rhino pregnancies.' But sadly none of these rhinos are northern whites.

Last Chance to Survive

When it became apparent that Najin and Fatu hadn't conceived, their keepers at Dvur Kralove were faced with a difficult decision. The rhinos could stay at the zoo and the IZW team could keep trying, or they could send their beloved rhinos somewhere else in the hope it would get their procreative juices flowing. 'It wasn't an easy choice,' says Stejskal. 'We had to do what was best for them.'

In the end, the zoo decided to send their last four fertile northern whites to Africa. Two females, Fatu and Najin, and two males, Sudan (Fatu's grandfather) and Suni, would be relocated under the banner of the 'Last Chance to Survive' project. In an ideal world, the animals would have been returned to Sudan or Chad or one of the countries where they used to live, but the world isn't ideal. It's certainly not safe for rhinos. In Africa, the rhino's main threat isn't habitat loss, but organised gangs of criminals who operate across borders. Their foot soldiers use high-tech equipment to locate the rhinos in the vast open grasslands, then tranquilise them, hack off their horns and let them bleed to death. Demand comes from Asia and the Middle East where rhino horn is carved to make ornamental daggers and ground down to make medicine. It is used to treat everything from gout to cancer, snakebites to demonic possession, but the irony is, of course, that it does no such thing. Rhino horn is made from keratin, the same substance that forms hair and fingernails, and has zero medicinal value. Retailing at over £41,000 (US$60,000) per kilo on the black market, it is more valuable than gold, diamonds or cocaine, making it the most expensive snake oil in the world.

With no safe havens left in the animals' former range, the decision was taken to send the Dvur Kralove rhinos to a maximum-security wildlife park in Kenya. The Ol Pejeta nature conservancy is 90,000 acres of rich, rolling savanna. A three-hour drive from Nairobi, it's nestled between the foothills of the Aberdares Mountains and Mount Kenya. It's home to all of Africa's 'Big Five': rhinos, elephants, lions, leopards and buffalo, which makes it attractive both to tourists and poachers. So its enormous electrified perimeter fence is patrolled round the clock by armed guards.

It seemed their best bet. And so, one cold, snowy morning shortly before Christmas 2009, the quartet were lured into crates and driven under police escort to Prague-Ruzyne airport, where they were loaded onto a jumbo jet bound for Nairobi. Twenty-six hours after their journey began the animals were unloaded under the blistering sun, apparently none the worse for wear, and took their first tentative steps on African soil.

Because the animals were so precious – the last four fertile animals of their kind – they were given extra layers of security. Their pen, or 'boma', was in the centre of the park, surrounded by an additional electrified fence, dotted with watchtowers and patrolled by dogs. Their horns were filed down and fitted with radio transmitters, to make them less desirable to poachers and easier to track. And they were given their own personal bodyguards to protect them round the clock. For Sudan, the only one of the animals to be born in the wild, it was a momentous return to the continent of his birth. For all of them, it was the start of an entirely new chapter in their lives.

The hope was that the natural surroundings would turn their thoughts to love and a few years later, romance did appear to be blossoming. Suni was seen mating with Najin, but as time wore on, it became clear that Najin had not become pregnant. So the keepers came up with another strategy. They arranged a 'blind date' between the northern white rhino females and a southern white rhino bull. It was a sensible idea. Back in the seventies, a hybrid calf that was half southern white, half northern white was born at Dvur Kralove. If Fatu or her mother became pregnant by a southern white rhino bull, then any resulting progeny would also be a hybrid. It might not be a genetically pure northern white, but at least the northern white rhinos' DNA would be preserved, albeit in a slightly diluted form. However the disappointment continued. There were no baby rhinos. To make matters worse, Suni was found dead in his enclosure in October 2014, apparently from natural causes. He was 34 years old.

At the end of that same year, Hildebrandt and his team visited Ol Pejeta to examine the remaining three animals, but what came next was an even more devastating blow. Sudan, they found, is too old and too weak to father children. His testicles have degenerated with age. Najin, his daughter, has problems with her hind legs and is unable to bear the weight of a pregnancy. Fatu, her daughter, has problems with her uterus. To top it all, genetic tests had previously revealed that both Sudan and Najin carried a genetic abnormality that could potentially interfere with their ability to reproduce. 'Our assessment was that neither of the females was capable of becoming pregnant anymore,' says veterinarian Robert Hermes, who works with Hildebrandt and examined the animals. 'We think the pathology that we found is not treatable.'

Dead Rhino Walking

The northern white rhino is not extinct … yet. But it's only a matter of time. Today, the only northern white rhinos left alive *anywhere* on the planet are the three animals,

Fatu, Najin and Sudan, that still live at Ol Pejeta. There is no possible way the animals can breed themselves out of trouble. They're too old, too ill and too related. Conservationists talk about species that have entered an 'extinction vortex', a downward spiral towards extinction from which they cannot naturally recover. That's where the northern white rhino is right now. It is 'functionally extinct'. Although a few stragglers remain, to all intents and purposes, the species has already gone. Fatu, Najin and Sudan are ghosts. They are the walking dead, a living embodiment of how human greed and carelessness is stripping species from the face of our planet.

The Ol Pejeta conservancy, the northern white rhinos' carers, and Hildebrandt and his team are now faced with an incredibly daunting task. How to save a subspecies that has just three infertile members left? And how to silence critics who tell them there is no point?

Rhinos are important, not just because they are big and beautiful, iconic and mesmerising, but because they help create and sculpt the landscape they live in. Like the passenger pigeon and woolly mammoth before them, rhinos are ecosystem engineers. White rhinos fashion and manicure the grasslands of the African savanna. Their box-shaped muzzles are perfect for close-cropping the grass they feed on, leading to the creation of lush, fertile 'grazing lawns' that other species like impala and wildebeest depend on. It's particularly important in high rainfall areas where the grasses grow taller and need more mowing. The rhinos' grazing helps create a patchy savanna that has both areas of long, non-grazed grass and short, nibbled lawn. This is important when the fires come. Every few years, fire rips across the savanna, but where there are rhinos, the grazing lawns they create act as natural firebreaks. As a result, the fires are smaller, more diffuse and return less frequently. Rhinos change the behaviour of fire.

Studies of the last Ice Age show us what happens when all the big grazers go. When the megaherbivores disappeared from Australia, mixed rainforest turned to scrubby bush. When they went from North America, the lush mammoth steppe was replaced by mossy tundra. There were cascading effects that trickled all the way from the top of the ecosystem right down to the bottom. If we lose the megaherbivores that we have today, we run the risk of creating an empty, barren landscape.

But there's more to it than that. Towards the end of the last Ice Age the world began to warm. It wasn't like a thermostat being turned up gradually; it was more like someone flicking a gas fire on and off repeatedly. Temperatures fluctuated. The ice sheets began to melt. Yet for a time at least, the plant and animal life remained more or less the same. It was only when the megafauna went extinct that the ecosystem began to change fundamentally. Paleoecologist Jacquelyn Gill from the University of Maine has studied this period of upheaval in great detail, and thinks the megaherbivores helped buffer against the effects of climate change. 'Modern and palaeontological studies suggest that when native large herbivores exist at sustainable levels, the ecosystems that they live in are more resistant to climate change and have greater levels of biodiversity,' says Gill. Now fast forward 12,000 years to the present day. Our climate is changing. The world is warming. If Gill is right, then the presence of rhinos and other big herbivores could help to enhance biodiversity and make ecosystems more resilient to this change. But they can't do it if they've all gone.

The northern white rhino is an animal that is still alive but that is in desperate need of de-extinction. When we think about de-extinction we are drawn instinctively to the species that are no longer with us, but there are species still alive whose future is so bleak that from a functional perspective, they might as well be gone. The Yangtze giant softshell turtle (*Rafetus swinhoei*), for example, is in a similar predicament. The largest freshwater turtle in the world, it was once common in parts of Southeast Asia, but just like the rhino, it was poached for use in Chinese medicine and its numbers have plummeted. Now there are just three of the snouted giants left: one in a protected lake in Vietnam and a pair at the Suzhou Zoo in China. They've laid eggs from time to time but none have been viable, and although artificial insemination has been trialled, it too has failed.

It makes sense to broaden our scope and include these species under the remit of de-extinction. From a practical point of view, de-extinction becomes more difficult the longer a species has been gone; ecosystems change and living relatives become more scarce. To choose an animal that is still alive – but almost beyond hope – makes perfect sense. With living members still present, we can study their biology and ecology, and amass the knowledge needed to de-extinct them while they are still here, rather than stabbing in the dark when they are gone.

If we don't intervene, it's inevitable that the northern white rhino will go extinct. Yet opinions are divided over what should happen next. Some think we should, in the words of the world's most annoying song, 'let it go', that it's already too late to do anything. But there are examples of species that *have* been sucked into an extinction vortex and made it out the other side. These vortices are not always the black holes they are painted to be, and with human ingenuity and science helping to pull a species back from the void, we massively increase the chances that a species will recover. Others argue that saving the northern white rhino is a distraction. Cash earmarked for conservation, they say, would be better spent on animals that can actually breed. 'We've spent money on the northern white rhino in the past,' says Susie Ellis, Executive Director of the International Rhino Foundation, 'but it's not our institutional focus anymore.' Some suggest that because the northern white rhino has a close living relative, the southern white rhino – which is 'only' listed as Near Threatened by the IUCN – it doesn't matter if the northern white rhino goes extinct. However, it does matter. It matters very much. People have argued for a long time about the relatedness of the northern and southern white forms. Some consider them separate 'subspecies', a blurry term lacking precise definition that tends to denote related animals that are genetically similar but geographically separate.* As such, they argue, the northern white rhino is not genetically distinct enough to be worthy of rescue. But a recent study that scrutinised the DNA of the two varieties found that they are as different from each other as they are from the black rhino.

We should de-extinct the northern white rhino because it is genetically distinct and ecologically valuable. A keystone species, it represents 'value for money' through the repercussions its presence has on its landscape and on the species with which it interacts. If we let the northern white rhino slip away because there are other rhinos still alive that look 'a bit like it', then what's to say we won't apply the same logic next time round … Let the Javan rhino go because it looks a 'bit like' the greater one-horned rhino. Let the Sumatran rhino go because it looks a 'bit like' the black rhino. Our big grazing herbivores are slipping away. Sixty per cent of them, including rhinos, elephants and gorillas, are at risk of extinction. It has to stop somewhere.

Nor should we be blasé about the fate of those rhinos that still occur in larger numbers. There may well be around 20,000 southern white rhinos and 5,000 black rhinos left in the wild, but recent years have seen record levels of poaching seriously reduce their populations. In 2015, poachers killed 1,315 rhinos in Africa, making it the deadliest year ever for the animal. Experts fear that soon, populations will reach a tipping point. The number of deaths will exceed the number of births, and after that, it doesn't take a genius to work out what follows. According

*The northern white is *Ceratotherium simum cottoni* while the southern white *Ceratotherium simum simum*.

to Save the Rhino, it's entirely possible that all of the world's rhinos will be extinct in the wild by 2026.

How Do You Like Your Eggs?

How to save the northern white rhino when its future seems so hopeless? In 2015, social media was briefly awash with photos of Sudan surrounded by his armed posse. The pictures were as beautiful as they were poignant. Against a wide African savanna and an expansive blue sky, a huge, imposing Sudan stood proudly in profile. All around him, as straight as pillars, stood his sentries clad in khaki, wearing military boots, clutching rifles. They stared towards the horizon, eyes on constant lookout for the first sign of trouble, while Sudan gently bowed his head. For a short time, because social media is fickle and transitory, Sudan became an internet sensation, the most famous rhino on the planet.

The media was keen to point out Sudan's poignant claim to fame, that of being the only living male northern white rhino left on Earth. The fate of an entire species, they said, rested on his shoulders. But it's not as straightforward as that. Already in his forties, Sudan is no spring chicken. He will die soon and when he does, he will be greatly missed. But the lack of his physical presence will have little bearing on the future of the northern white rhino. Thanks to Hildebrandt's electro-ejaculator, semen samples from Sudan and several other males have already been collected and are now stored in cryo-banks waiting to be used. 'Everyone thinks the crisis is the last male,' says Richard Vigne, Chief Executive Officer at Ol Pejeta, 'but we have lots of sperm stored. The bigger crisis is the lack of females and their eggs.' Fatu and Najin, the last two female northern white rhinos on Earth, are the ones who really deserve the media spotlight.

Conservationists, cell biologists and other interested parties meet regularly to debate the best way to save the northern white rhino, and there are options. Unfortunately, every single one of them relies on being able to source and store northern white rhino eggs. For years this task was thought impossible. For a start, it can be very difficult to tell when a female northern white rhino is ovulating. Some rhinos – greater one-horned and Sumatran – 'sing' when they ovulate. They make peculiar vocalisations not heard at other times of the month. But white rhinos make no such fuss, and although male northern whites may be able to sense the hormonal changes that occur around ovulation, they are not obvious to our human senses. To physically 'go in' and retrieve an egg is also problematic. Not only is the female's genital tract over a metre (3.3 feet) long, it is also, I am reliably informed, a 'tortuous structure', full of 90-degree turns and dead ends. This puts a female's ovaries and eggs beyond the reach of all but the most specialist laparoscopic equipment. Then there's the problem of storage. Unless an egg is to be used immediately, it needs to be preserved. But freezing is difficult because the large egg cells shatter, and other storage methods have also proved problematic. So for a while it was presumed almost impossible to store rhino eggs.

Thomas Hildebrandt and his team, however, have spent the last 15 years refining the techniques needed to overcome these problems. 'My philosophy has been that we work step by step, optimising one procedure before moving on to the next,' says Hildebrandt. The methods they have developed are a mix of high-tech interventions and, where possible, standing back to let nature run its course.

To tell when a female rhino is ovulating, Hildebrandt has enlisted the help of those who care for the rhinos the most: their keepers. At Ol Pejeta, Dvur Kralove and elsewhere, the keepers form very close bonds with the rhinos they tend. They don't just feed them, wash them and muck them out. They talk to them softly, tickle them behind the ears and rub their bellies with a yard brush. In return, they have the rhinos' trust and are able to monitor any physical changes that occur. Working with the keepers,

Hildebrandt has discovered that one or two days before a white rhino ovulates, a plug of mucus is released from the vagina. It's the animal's way of having a bit of a clear out, sprucing things up and getting rid of any dead cells, so that the female is ready for sex. By recording when the discharge is released, Hildebrandt and the IZW team are able to calculate the length of each individual female's menstrual cycle, and from that predict exactly when a female will next ovulate. It's incredibly accurate – so much so, that the group use the strategy to pre-order their airline tickets.

When I track down Hildebrandt to talk about his work, he's just back from the Czech Republic, where he has cracked the next step: collecting rhino eggs. With the rhino's internal plumbing more convoluted than the plot of a Scandi-noir box set, Hildebrandt has devised an alternative route to reach the ovary. Under general anaesthesia,* he guides a two-metre (6.6 feet) long, specially designed, patent-pending egg-retrieval device up the animal's rectum, then punctures a hole through to the ovary on the other side. This enables him to aspirate and retrieve the eggs. It might sound crude, but with care, precision and an astute dose of antibiotics, the risk of infection and complications is minimised. Hildebrandt and co. recently used the method to harvest five eggs from a southern white rhino female at Dvur Kralove Zoo. The eggs, however, are not fully mature, so the next step is to bathe them in a carefully concocted mix of nutrients in a culture dish. It's a delicate process that has been optimised by Cesare Galli and Giovanna Lazzari of Avantea, an Italian biotech company that is now collaborating with the IZW. Critically, it causes the eggs to ripen, and then and only then can they be either frozen or used to make a baby rhino.

Collaborating closely with Fatu and Najin's keepers, the IZW team now plans to make twice-yearly trips to Ol Pejeta to collect eggs from the females for as long as

*The rhino, not Hildebrandt.

they continue to make them. The menstrual cycle of a white rhino lasts between 30 and 35 days, so the animals have plenty of time to recover and cycle normally between visits. 'We have everything in place,' he says. The first tranche of eggs will be matured then frozen as an insurance policy, but after that, Hildebrandt will move into unchartered territory. Having tried and unfortunately failed to make northern white rhinos via artificial insemination, where semen is introduced directly into the female's reproductive tract, the team will now try the next best option: to make a 'test-tube rhino' and transfer that back into a surrogate womb.

In 2015, they took a massive step towards this goal when Avantea scientists managed to harvest four eggs from the body of Fatu's then recently deceased aunt, Nabire. Thinking it would give them the best chance of success, they decided to try fertilising the eggs with sperm from a southern white rhino and, much to their delight, managed to create one little IVF embryo. Although it stopped developing when it was still a tiny bundle of cells, the development buoyed the spirits of those trying to save the northern white. 'This is the first time anyone has managed to fertilise a northern white rhino oocyte [egg] in a dish,' says Hildebrandt. 'It's a big achievement.' And he's right. It might not be the northern white rhino calf he so desperately hopes for, but it's the nature of cell biology research to proceed in small incremental steps. It would be welcome but entirely unrealistic to expect early attempts at rhino IVF to produce a live calf. In the meantime, however, while Hildebrandt and his collaborators continue to work methodically, polish their techniques and make the best of the limited resources they have, Fatu and Najin's biological clocks are ticking.

The problem, explains team member Robert Hermes, is that the IZW team is having to refine its protocols with limited biological resources of variable quality. 'If I worked with dairy cows,' he says, 'I could go to the abattoir and get

100 cattle ... 200 ovaries ... and then collect many hundreds of eggs.' It would provide enough basic research material to try lots of different protocols and tease out what works best, for the cow at least. But rhino protocols need to be devised with rhino eggs and, like the rhinos they come from, those are in short supply. It's for this reason that the team has been refining its protocols on less endangered rhino species such as the black and southern white rhino, but even then there are problems. Around 50 per cent of rhinos in captivity, the stock from which Hermes and Hildebrandt draw, have reproductive problems of some sort. So although they may still be releasing eggs, the quality of those eggs is uncertain. If only there was some other source of northern white rhino sperm and eggs ...

The Y Factor

Many years ago, when I worked as a cell biologist in the laboratory at London's Institute of Psychiatry, we laboured under two major misunderstandings; first, that regular, almost constant tea breaks would greatly improve scientific productivity, and second, that making stem cells was problematic. At the time, people were very excited about making stem cells because of their promise for medical research. But how to get hold of them? Back then, the only way to source genuine human stem cells was to take an IVF-created embryo, just a few days after conception when it was still a tiny ball of cells, and use a fine needle to aspirate the stem cells from it. But there were lots of people who found the idea unpalatable because the embryo was destroyed in the process. It didn't matter that these were 'surplus' embryos never destined for the comfort of a mother's womb; the pro-life lobby still hated it. 'Making' stem cells was not just difficult, it was ethically charged.

Then along came a man called Shinya Yamanaka from Japan's Kyoto University, who not only solved the moral conundrum, but also came up with a way of making stem cells that was easy. He made them from skin cells, without the need for embryos at all. In 2006, Yamanaka took skin cells from an adult mouse, added a handful of genes and reprogrammed the DNA inside the skin cells into a stem cell-like state. The cells that he produced, which he called 'induced pluripotent stem cells', or 'iPS' cells for short, could then be coaxed to become other cell types, including nerve and heart cells, proving that his iPS cells behaved like stem cells. It was cellular alchemy; Nobel Prize-winning gold. Ian Wilmut, creator of Dolly the sheep, then at the MRC Centre for Regenerative Medicine in Edinburgh, described it as 'one of the big experiments of the decade, maybe even the century.' In no time at all, other laboratories began to adopt Yamanaka's treatment. They added the same four genes – or 'Yamanaka factors', as they've become known – to adult cells and began to make their own iPS cells. And it wasn't just mouse skin cells that could be given the iPS makeover. Scientists soon showed that iPS cells could be generated from the cells of different species, including rats, monkeys and humans. Meanwhile, elsewhere, in separate experiments, scientists have shown how stem cell-derived eggs and sperm can be used to create healthy, live mice. The studies raise the prospect that in the future, scientists might be able to take a skin cell, turn it into a stem cell, then use that cell to generate eggs and sperm to help generate test-tube babies for infertile couples. Intriguingly, it also raised the possibility that if this could be done for humans, it could be done for other species, too, including endangered ones.

It was something that stem cell researcher Jeanne Loring from the Scripps Institute in La Jolla, California, was thinking about when she was looking for an excuse to take the researchers in her laboratory on a day out. 'I wanted to take them to the San Diego Zoo,' she says. 'It's one of my favourite places, but we needed a reason to be there.' So they decided to meet with Oliver Ryder, the Director of Genetics there, and discuss whether stem cells could somehow be used in wildlife conservation. To Loring, it was obvious. Skin or other cells collected from endangered animals could, in theory, be reprogrammed to become iPS cells, which could then be used to create eggs and sperm, which could then be used to make babies. Loring, who had decades of experience working with stem cells, had the knowhow, and Ryder had the raw material.

A little over 40 years ago, Ryder and colleagues began collecting skin samples from rare and endangered animals in the hope that one day, they might come in useful. Over the years, the collection, known as the Frozen Zoo, has grown into one of the biggest cell banks in the world.

Squirreled away in its many bubbling vats of liquid nitrogen are more than 70,000 samples from over 700 different species of mammals, birds, reptiles, amphibians and fish. The cells have been used in hundreds of scientific studies, helping researchers understand how species evolve and genetic variation changes over time. When the black-footed ferret had its genome sequenced, they used cells from the Frozen Zoo. Critically, for the future of Fatu and her kind, Ryder has been stockpiling cells from the northern white rhino, starting with a female called Lucy, whose cells were frozen in 1979. Since then, the Zoo has amassed a veritable 'crash' of northern white rhino cells, including samples from Fatu donated by the IZW team.* Frozen in tiny vials are the cells of 12 different animals: eight unrelated individuals and four of their offspring. According to Ryder, that amounts to a significant sampling of genetic diversity and, if used wisely, should be enough to help save the subspecies from extinction. There is more of a gene pool preserved in these tiny tubes than survives in all of the remaining live animals put together.

*A 'crash' is the collective noun for rhinos. Other African animal collectives include a leap of leopards, a cackle of hyenas and my favourite, a dazzle of zebras.

A short while after their trip to the San Diego Zoo, one of Loring's post-docs, Inbar Friedrich Ben-Nun, thawed some of Fatu's cells and began to grow them in a dish. Adding a judicious dash of the seemingly magical 'Y' factors, the cells were turned into their more versatile stem cell state. Loring and her team had made the world's first iPS cells from an endangered species, and the signs are that the cells are every bit as talented as the team had hoped. In a tissue culture setting, they can divide and be coaxed to become many different types of cell. The next step, says Loring, is to try to turn them into rhino eggs and sperm.

Researchers haven't given up on the idea of using the eggs that are directly harvested from Najin or Fatu, rather they are aware that they need a Plan B. 'We realise now that to save the northern white rhino, we are going to have to use some advanced cellular techniques, such as iPS cells,' says Dvur Kralove Zoo's Jan Stejskal. And with iPS cells come new options.

If fully functioning rhino sperm and rhino eggs can be made from iPS cells, they could be 'married' together in a dish and used to make embryos. With samples of their skin cells already in the Frozen Zoo, there'd be no need to subject Najin and Fatu to the twice-yearly egg collections, and Loring's team now hope to make iPS cells from the other 11 northern white rhinos whose skin cells have been frozen. If the technology is reliable, it would offer researchers an almost limitless supply of northern white rhino eggs and sperm with which to optimise their methods, and would allow them to mix and match genomes of different animals at will, maximising that all important genetic diversity.

The resultant embryos would be nurtured in a dish for a few days then transferred to a surrogate's womb. Because of their health problems, Fatu and Najin would not be used as surrogates. Instead the world's next northern white rhino will be born to its cousin, the southern white. At Ol Pejeta, they have already earmarked four young, fertile females for the job. They live in a huge pasture, in large social groups; important, Hildebrandt believes, because natural behaviour and social interaction are key to keeping rhinos both happy and fertile. Before the embryo is implanted, the surrogate will be allowed one final fling with a sterilised bull rhino because it's thought his sperm-free ejaculate will help make her uterus more receptive to the embryo. It will prime her body for pregnancy and hopefully give the embryo a better chance of survival. And while none of this has ever been attempted before, Hildebrandt remains unfazed. 'We have successfully transferred embryos back in many other mammal species,' he says. 'The rhino reproductive tract may well be in a class of its own, but I am optimistic that we will master this step in the near future.'

Little by little, all the pieces of technology needed to save the northern white rhino are dropping into place, and iPS cells, so versatile and easy to make, could represent a game-changer for the fate of this larger-than-life iconic wonder. But it's not just the northern white that could benefit from this technology.

Loring's team has also made iPS cells from the drill (*Mandrillus leucophaeus*), a short-tailed mandrill-like monkey whose numbers have more than halved in the last 30 years. In Thailand, Anucha Sathanawongs at Chiang Mai University has made elephant iPS cells. The immediate goal is to devise therapies for the country's ailing captive elephants, many of whom suffer from ulcers and arthritis, but Sathanawongs is also hoping to coax the elephant iPS cells into eggs and sperm. If the technique works and can be replicated elsewhere, it could provide a ready supply of eggs for those interested in bringing back the woolly mammoth, or an approximation of it. iPS cells have been made from the endangered snow leopard (*Panthera uncia*), from chimpanzees (*Pan troglodytes*), gorillas (*Gorilla gorilla*) and orangutans (genus *Pongo*). The development of iPS cells means that repositories such as the Frozen Zoo are now more important than ever. They're not just preserving the cells and genomes of the world's wildlife, they are a starting point for new life, a vital repository of genetic diversity which, if used carefully, has the potential to revive not just individual animals, but entire, viable populations. We shouldn't just be trying to save wildlife: where possible, we should be saving their cells, too.

Bring Back the King

When I first heard about de-extinction, I was intrigued but uncertain. I have since determined that I am fully in favour of scientists developing new techniques, not just to ease the pace of extinction, but to also reverse it. We need to proceed cautiously. We need to ensure these options are safe, effective and have public buy-in, but they offer a vast amount of potential that deserves at least to be explored. I've realised that if we are to bring a creature back from the past then we had best choose one that we know a lot about, one whose reproductive physiology is known, whose cells we can manipulate. It had best be an animal that we love, that we have space for, that we miss and we mourn. It should be a creature whose ecology is well known, a keystone species whose return would have a positive ecological impact. It should be an animal that warms the cockles, that could inspire future generations to look after their planet and prevent future extinctions from happening.

Everyone has heard about the current biodiversity crisis. We all know that flora and fauna are in trouble, but because the vast majority of species are going extinct quietly, invisibly, in places hidden from our everyday lives, it's all too easy to become sucked into the misguided belief that extinction is not happening on our watch. Species used to go extinct in the past, species will go extinct in the future, but today, as I sit in my centrally heated home, sipping my coffee, scanning the internet, nothing has changed. In the time it has taken you to read this chapter, another species has disappeared. In the time it's taken me to write this book, *at least* 20,000 species have gone extinct, but save for the odd one that makes the official IUCN Red List most of them are completely unknown. We live in the midst of the sixth mass extinction, but it's largely invisible or ignored.

When I first started thinking about de-extinction, I asked myself which *single* creature would I most like to see brought back from extinction? In the real world, of course, we're not restricted in this way. I may not have chosen the woolly mammoth or the gastric-brooding frog or the passenger pigeon to top my list, but I wish the researchers working to bring them back the very best of British. There are good, sound reasons for bringing these particular creatures back. When the first healthy de-extinct animal is born, it will be a truly incredible achievement. I look forward to that day. But if I had to choose just one ...?

As I write these final pages, I realise I have arrived at an answer that I never expected. The animal I would most like to bring back from extinction is one that is still alive ... just. Because of our greed and our lack of respect for nature, the northern white rhino is nearly gone. Sudan, the last male standing, is entering his twilight years. One eye is clouded by cataracts. When he's around other rhinos, they bully him. It makes him lose condition, so his carers at Ol Pejeta have created a new enclosure just for him. Fatu and Najin live next door and although he is unable to see them, he can smell them and knows they are there. 'To create a northern white rhino before the last one disappears is the driving force for me,' says Hildebrandt. Fatu, the youngest of the three, may well be that last animal. 'We hope that Fatu at least will get to meet a new member of her own kind,' he says. The only thing that can save the northern white rhino now is science. If I had to choose just one animal to de-extinct, then the wrinkly-faced, twinkly-eyed northern white rhino would be it.

We are responsible for the downfall of the northern white rhino, but with assisted reproduction, stem cell science and genome editing, we have at our fingertips some of the most powerful technology on the planet. It's a science that can save species, shape evolution and sculpt the future of life on Earth. We are so close to being able to use these methods to make facsimiles of species that are long gone, to enhance the genomes of animals that are still with us and offer a lifeline to species that are on the brink. It's up to us to decide how we use this knowledge. I believe it was Spiderman's great uncle, the late, great Uncle Ben, who summed it up best when he turned to his web-spinning nephew and said, 'Remember, with great power comes great responsibility.'

As I write this line, there are just three northern white rhinos left alive on the planet. By the time you read it, they might all be gone.

P.S.

So that's it.

The end.

Finished.

Elvis has left the building.

You can go home now…

…

What's that?

…

You're already home. You've finished the book but you're still thinking about dodos and dinosaurs and de-extinction, and what a mess the planet is in and how you wish that Sudan and Najin and Fatu weren't going to die and how you hope that scientists will somehow save the northern white rhino and all the other animals that need our help…

I know.

Me too.

Big hugs.

If you're interested and feel that you would like to, you can donate to the work of the Leibniz Institute for Zoo and Wildlife Research at: www.izw-berlin.de/SaveTheNorthern WhiteRhino.html. The money will go directly towards their work developing assisted reproductive techniques for endangered species, including the northern white rhino. But if passenger pigeons and black-footed ferrets are more your bag, and you feel that you would like to, you can donate to the work of Revive and Restore at www.reviverestore.org/ No pressure. You don't have to. Just thought I'd mention it.

Key References
A Little Less Conversation, a Little More Reading

This is not an exhaustive list of references, rather a bijou bibliography of websites, papers and articles that I have found helpful. If you want to find out more about de-extinction, the best place to start is the website of Revive and Restore: reviverestore.org. *It's a brilliant resource – up-to-date, reader friendly, packed with info about ongoing projects and technology. It even has an Extinct Species Colouring Sheet.*

Introduction: Bringin' It Back

The first de-extinction: Folch, J. *et al.* First birth of an animal from an extinct subspecies (*Capra pyrenaica pyrenaica*) by cloning. *Theriogenology* 71 (6): 1026-34 (2009).

It inspired Jurassic Park, *the discovery of cellular nuclei inside an amber-entombed fly:* Poinar G. O. & Hess, R. Ultrastructure of 40-Million-Year-Old Insect Tissue. *Science* 215 (4537): 1241-1242 (1982).

These two papers are depressing but well worth the read. They describe the alarming state and rate of present-day extinctions: (1) Dirzo, R. *et al.* Defaunation in the Anthropocene. *Science* 345 (6195): 401-406 (2014). (2) Urban, M. C. Accelerating extinction risk from climate change. *Science* 348 (6234): 571-573 (2015).

In 2013, there was a TEDx event all about de-extinction. Organised by Stewart Brand and Ryan Phelan from Revive and Restore – you can watch it here: www.longnow.org/revive/events/tedxdeextinction/the-program/

Chapter 1: King of the Dinosaurs

The paper that kicked it all off. Mary Schweitzer discovers structures that 'look like' red blood cells inside fossilised dinosaur bone, prompting speculation that organic matter can survive the fossilisation process: Schweitzer M. H. *et al.* Haem compounds in dinosaur trabecular bone. *PNAS* 94 (12): 6291-6296 (1997).

Critics attempt to pan Schweitzer's work: Pevzner, P. A. *et al.* Comment on 'Protein Sequences from Mastodon and *Tyrannosaurus rex* Revealed by Mass Spectrometry.' *Science* 321 (3892): 1040 (2008).

But it seems that organic matter may survive the fossilisation process more often than we think: Bertazzo, S. *et al.* Fibres and cellular structures preserved in 75-million-year-old dinosaur specimens. *Nature Communications* 6: 7352 (2015).

The first ancient DNA to be recovered: Higuchi, R. *et al.* DNA sequences from the quagga, an extinct member of the horse family. *Nature* 312: 282-284 (1984).

And the prize for the oldest DNA discovered so far goes to … a horse: Orlando L. *et al.* Recalibrating *Equus* evolution using the genome sequence of an early Middle Pleistocene horse. *Nature* 499: 74-78 (2013).

Has Schweitzer found evidence for DNA preservation in a T. rex fossil? The jury is out: Schweitzer, M. H. *et al.* Molecular analyses of dinosaur osteocytes support the presence of endogenous molecules. *Bone* 52 (1): 414-423 (2013).

Jack Horner's plans to build a dinosaur revealed in: Horner, J. & Gorman, J. 2010. *How to Build a Dinosaur: The New Science of Reverse Evolution.* Plume, New York City.

Not quite a dinosaur but a seriously ugly bird. Scientists create a chicken embryo with a snout: Bhullar, B-A. S. *et al.* A molecular mechanism for the origin of a key evolutionary innovation, the bird beak and palate, revealed by an integrative approach to major transitions in vertebrate history. *Evolution* 69 (7): 1665-1677 (2015).

Chapter 2: King of the Cavemen

Svante Pääbo is the first to retrieve DNA from a Neanderthal: Krings M. *et al.* Neanderthal DNA Sequences and the Origins of Modern Humans. *Cell* 90: 19-30 (1997).

Svante Pääbo is an ingenious, talented and inspirational scientist. No one tells his story better than the man himself: Pääbo, S. 2014.

Neanderthal Man: In Search of Lost Genomes. Basic Books, New York.

Hendrik Poinar extracts nuclear DNA from fossil sloth poo: Poinar, H. Nuclear gene sequences from a late Pleistocene sloth coprolite. *Current Biology* 13 (13): 1150-1152 (2003).

It's a draft but apparently that's OK. Green, R. E. *et al.* A draft sequence of the Neanderthal genome. *Science* 328 (5979): 710-722 (2010).

The New York Times *reports that a Neanderthal could be made for US$30million:* Feb 12, 2009. www.nytimes.com/2009/02/13/science/13neanderthal.html

Hugely entertaining: Wynn, T. & Coolidge, Frederick L. 2012. *How to Think Like a Neanderthal.* Oxford University Press, Oxford.

Evidence that Neanderthals and modern humans interbred: Vernot, B. & Akey, J. M. Resurrecting Surviving Neanderthal Lineages from Modern Human Genomes. *Science* 343 (6174): 1017-1021 (2014).

Chapter 3: King of the Ice Age

An insight into the mind and works of one of the world's most brilliant geneticists. George Church has speculated about de-extincting a Neanderthal and is busy making a cold-loving elephant: Church, G. & Regis, E. 2012. *Regenesis: How Synthetic Biology Will Reinvent Nature and Ourselves.* Basic Books, New York.

The early days of mammoth de-extinction: Stone, R. 2003. *Mammoth: The Resurrection of an Ice Age Giant.* Fourth Estate, London.

That bloody mammoth: Grigoriev, S. *et al.* Discovery of a woolly mammoth (*Mammuthus primigenius*) carcass from Malyi Lyakhovski Island (New Siberian Islands). *Scientific Annals of the School of Geology*, Aristotle University of Thessaloniki 102: 64-76 (2014).

I'm always finding things that I'd forgotten about at the bottom of my deep freeze. Here's what to do if you find a dead mouse in yours: Wakayama, S. *et al.* Production of healthy cloned mice from bodies frozen at -20°C for 16 years. *PNAS* 105 (45): 17318-17322 (2008).

A load of old bull. Japanese scientists clone livestock from frozen testicles: Hoshino, Y. *et al.* Resurrection of a Bull by Cloning from Organs Frozen without Cryoprotectant in a -80°C Freezer for a Decade. *PLoS One* 4 (1): e4142 (2009).

Researchers inject mammoth nuclei into mice eggs but not a squeak: Kato, H. *et al.* Recovery of cell nuclei from 15,000 years old mammoth tissues and its injection into mouse enucleated matured oocytes. *Proceedings of the Japan Academy Series B Physical and Biological Sciences* 85 (7): 240-247 (2009).

Scientists may not have de-extincted the woolly mammoth, but they have de-extincted one of its proteins: Campbell, K. L. *et al.* Substitutions in woolly mammoth haemoglobin confer biochemical properties adaptive for cold tolerance. *Nature Genetics* 42: 536-540 (2010).

The woolly mammoth genome: Lynch,V. J., *et al.* Elephantid genomes reveal the molecular bases of Woolly Mammoth adaptations to the arctic. *Cell Reports* 12 (2): 217–218 (2015).

It's the stuff that patent battles are made of – Jennifer Doudna and Emmanuelle Charpentier use CRISPR to cut the genome with extreme precision: Jinek, M. *et al.* A programmable dual-RNA-guided DNA endonuclease in adaptive bacterial immunity. *Science* 337 (6096): 816-821 (2012).

No one asks the paleoecologists what they think about bringing Ice Age animals back, but they should. Jacquelyn Gill has some eloquent thoughts on de-extincting the woolly mammoth in her blog: https://jacquelyngill.wordpress.com/

Chapter 4: King of the Birds

Everything you ever need to know about dodo history: Fuller, E. 2002. *Dodo: From Extinction to Icon.* HarperCollins, London, New York.

From historical records, the date of the dodo's demise can be calculated: Hume, J. P. *et al.* Palaeobiology: Dutch diaries and the demise of the dodo. *Nature* 429 (2004).

DNA analysis reveals that the dodo is officially a pigeon: Shapiro, B. *et al.* Flight of the Dodo. *Science* 295 (5560): 1683 (2002).

Recent excavations in Mauritius are revealing more about the dodo and its ecology: Hume, J. P. *The Dodo: from extinction to the fossil record. Geology Today* 28 (4): 147-151 (2012).

The Great Passenger Pigeon Comeback, everything you need to know: www.longnow.org/revive/projects/the-great-passenger-pigeon-comeback/

Everything you ever wanted to know about the passenger pigeon: Fuller, E. 2015. *The Passenger Pigeon.* Princeton University Press, New Jersey.

DNA tickled from passenger pigeon toe pads: Fulton, T. L. *et al.* Case study: recovery of ancient nuclear DNA from toe pads of the extinct passenger pigeon. *Methods in Molecular Biology* 840: 29-35 (2012).

Thinking about cloning a clucker? Then you need to know how to spot an egg's DNA against its blobby, yellow background: Kjelland, M. E. *et al.* Avian cloning: Adaptation of a technique for enucleation of the avian ovum. *Avian Biology Research* 7 (3): 131-138 (2014).

A duck fathers a chicken ... honestly, it really did: Liu, C. *et al.* Production of Chicken Progeny (*Gallus gallus domesticus*) from Interspecies Germline Chimeric Duck (*Anas domesticus*) by Primordial Germ Cell Transfer. Biology of Reproduction 86 (4): 101, 1–8 (2012).

Beautifully written, well worth a read: Avery, M. 2014. *A Message From Martha: The Extinction of the Passenger Pigeon and Its Relevance Today.* Bloomsbury, London, New York.

Chapter 5: King of Down Under

Paddle, R. 2000. *The Last Tasmanian Tiger: The History and Extinction of the Thylacine.* Cambridge Press, Cambridge.

Fuller, E. 2013. *Lost Animals: Extinction and the Photographic Record.* Bloomsbury, London, New York.

Schuster, S. C. *et al.* The mitochondrial genome sequence of the Tasmanian tiger (*Thylacinus cynocephalus*). *Genome Research* 19 (2): 213-220 (2009).

Thylacine DNA lives again: Pask, A. J. *et al.* Resurrection of DNA function in vivo from an extinct genome. *PLoS One* 32 3 (5): e2240 (2008).

On the global frogocalypse: Skerratt, L. F. *et al.* Spread of chytridiomycosis has caused the rapid global decline and extinction of frogs. *EcoHealth* 4: 125 (2007).

A frog cloning classic: Briggs, R. & King, T. J. Transplantation of living nuclei from blastula cells into enucleated frogs' eggs. *PNAS* 38 (5): 455-463 (1952).

Chapter 6: King of Rock 'n' Roll

If you want him, come sign here … the glorious, the one and only 'Americans for Cloning Elvis': americansforcloningelvis.bobmeyer99.com/

Find out more about Joni Mabe's Panoramic Encyclopaedia of Everything Elvis at: www.roadsideamerica.com/story/16788

Hairy genomes … the first nuclear genome recovered from ancient hair: Rasmussen, M. *et al.* Ancient human genome sequence of an extinct Palaeo-Eskimo. *Nature* 463: 757-762 (2010).

The 1000 Genomes Project: http://www.1000genomes.org/

A classic twins study: Bouchard, T. J. *et al.* Sources of Human Psychological Differences: The Minnesota Study of Twins Reared Apart. *Science* 250 (4978): 223-228 (1990).

Fifty years of nature versus nature: Polderman, T. J. C. *et al.* Meta-analysis of the heritability of human traits based on fifty years of twin studies. *Nature Genetics* 47: 702-709 (2015).

A genetic component to musical ability: Vinkhuyzen, A. A. *et al.* The Heritability of Aptitude and Exceptional Talent Across Different Domains in Adolescents and Young Adults. *Behavior Genetics* 39 (4): 380-392 (2009).

How is it that identical twins end up different? A piece that I wrote for New Scientist: Pilcher, H. The third factor: Beyond nature and nurture. *New Scientist*, 28 August (2013).

Poor parenting influences epigenetics influences how your pups turn out: Weaver, I. C. G. *et al.* Epigenetic reprogramming by maternal behaviour. *Nature Neuroscience* 7: 847-854 (2004).

This study shows how identical twins can sometimes become more different epigenetically with time: Talens, R. P. *et al.* Epigenetic variation during the adult lifespan: cross-sectional and longitudinal data on monozygotic twin pairs. *Aging Cell* 11 (4): 694-703 (2012).

The epigenomes of naturally and artificially conceived twins are different: Loke, Y. J. *et al.* Association of in vitro fertilisation with global and IGF2/H19 methylation variation in newborn twins. *Journal of Developmental Origins of Health and Disease* 6 (2): 115-124 (2015).

Why, if you made an infinite number of Elvis's and raised them all in the same Tupelo shack, they'd still turn out different every time: Freund, J. *et al.* Emergence of Individuality in Genetically Identical Mice. *Science* 340 (6133): 756-759 (2013).

Scientists use CRISPR to edit the genomes of human embryos: Liang, P. *et al.* CRISPR/Cas9-mediated gene editing in human tripronuclear zygotes. *Protein & Cell* 6 (5): 363-372 (2015).

Chapter 7: Blue Christmas

Conservationist Phil Seddon and colleagues discuss how best to select candidates for de-extinction: Seddon, P.J. *et al.* Reintroducing resurrected species: selecting DeExtinction candidates. *Trends in Ecology & Evolution* 29 (3): 140-147 (2014).

Scientists revive a 30,000-year-old Siberian virus: Legendre, M. *et al.* Thirty-thousand-year-old distant relative of giant icosahedral DNA viruses with a pandoravirus morphology. *PNAS* 111 (11): 4274-4279 (2014).

Does a woolly mammoth need a lawyer?: Carlin, N. *et al.* How to Permit Your Mammoth: Some Legal Implications of 'De-Extinction'. *Stanford Environmental Law Journal* (2014). stanford.io/1SBEOKE

The sad, sad story of the Yangtze River Dolphin: Turvey, S. 2009. *Witness to Extinction: How we Failed to Save the Yangtze River Dolphin*. Oxford University Press, Oxford.

Christmas Island rats wiped out by infectious disease: Wyatt, K. B. *et al.* Historical mammal extinction on Christmas Island (Indian Ocean) correlates with introduced infectious disease. *PLoS One* 3 (11): e3602 (2008).

The United States National Park Service on Yellowstone's gray wolf re-introduction: www.nps.gov/yell/learn/nature/wolf-restoration.htm

Chapter 8: I Just Can't Help Believing

How to grow the sperm of an endangered bird inside a chicken: Wernery, U. *et al.* Primordial germ cell-mediated chimera technology produces viable pure-line Houbara bustard offspring: potential for repopulating an endangered species. *PLoS One* 29: 5 (12): e15824 (2010).

Scientists have successfully cloned the endangered Esfahan mouflon. This paper reports on some of their early work: Hajian, M. *et al.* 'Conservation cloning' of vulnerable Esfahan mouflon: in vitro and in vivo studies. European Journal of Wildlife *Research* 57 (4): 959-969 (2011).

How to save the black-footed ferret?: Wisely, S. M. *et al.* A Road Map for 21st Century Genetic Restoration: Gene Pool Enrichment of the Black-Footed Ferret. *Journal of Heredity* 106 (5): 581-92 (2015).

Cataloguing genetic variation in living tigers and museum specimens: Mondol, S. *et al.* Demographic loss, genetic structure and the conservation implications for Indian tigers. *Proceedings of the Royal Society B* 280: 20130496 (2013).

Genetic methods could help save endangered species: Thomas, M. A. *et al.* Ecology: Gene tweaking for conservation. *Nature* 501 (7468): 485-486 (2013).

Some like it hot; scientists reveal genes that may help trout survive in warmer waters: Rebl, A. *et al.* Transcriptome profiling of gill tissue in regionally bred and globally farmed rainbow trout strains reveals different strategies for coping with thermal stress. *Marine Biotechnology (NY)* 15 (4): 445-460 (2013).

In the interest of balance, here's Ben Minteer on the ethics of de-extinction: Minteer, B. Is it right to reverse extinction? *Nature* 509 (7500): 261 (2014).

Chapter 9: Now You See It

For the latest on the northern white rhino, head to the Ol Pejeta Nature Conservancy's website: http://www.olpejetaconservancy.org/

Rhinos are ecosystem engineers: Waldram, M. S. *et al.* Ecological Engineering by a Mega-Grazer:White Rhino Impacts on a South African Savannah. *Ecosystems* 11: 101-112 (2008).

It's not looking good for the world's megaherbivores: Ripple, W. J. *et al.* Collapse of the world's largest herbivores. *Science Advances* 1 (4), e1400103 (2015).

First attempts at rhino IVF: Hermes, R. *et al.* Ovarian superstimulation, transrectal ultrasound-guided oocyte recovery, and IVF in rhinoceros. *Theriogenology* 72 (7): 959-968 (2009).

How to persuade a rhino to ovulate: Hermes, R. *et al.* Estrus induction in white rhinoceros. *Theriogenology* 78 (6): 1217-1223 (2012).

Artificial insemination in rhinos: Hermes, R. *et al.* First successful artificial insemination with frozen-thawed semen in rhinoceros. *Theriogenology* 71 (3): 393-399 (2009).

Cellular alchemy, Shinya Yamanaka makes stem cells from skin cells: Takahashi, K. & Yamanaka, S. Induction of pluripotent stem cells from mouse embryonic and adult fibroblast cultures by defined factors. *Cell* 126 (4): 663-676 (2006).

Stem cells make eggs make live mice: Hayashi, K. *et al.* Offspring from oocytes derived from in vitro primordial germ cell-like cells in mice. *Science* 338 (6109): 971-975 (2012).

A crash of rhino iPS cells are made: Ben-Nun, I. F. *et al.* Induced pluripotent stem cells from highly endangered species. *Nature Methods* 8: 829-831 (2011).

Acknowledgements
The Wonder of You

A few years ago, one of my best friends, the jaw-droppingly awesome Toni Harrington, asked me if I would do 'something' for Nottingham's first ever children's book festival, Tales from the Riverbank. She promised to buy me beer, and so, a short while later, I found myself in a suburban church hall performing a hastily concocted, fossil-fuelled romp called 'Bring Back the Dinosaurs' to a spirited eruption of schoolchildren.[*] They were feisty, they were scary, they were oh so very lairy (the children, not the dinosaurs). Never have I met an audience *so* determined to join in! Somehow, I managed to survive the experience and make it out of Nottingham without going the way of the dodo. Then, over time, as the emotional (and physical) scars from my feistiest gig ever began to fade, 'Bring Back the Dinosaurs': The Show slowly evolved into *Bring Back the King*: The Book. I hope you have enjoyed it.

The book may be nearly finished, but before I'm done, I would like to take the opportunity to thank all those who have lent a hand along the way. First and foremost, a huge and sincere thank you to all of the animals that have ever gone extinct. Without you, there would be no *Bring Back the King*. It's not without irony that I note how your selfless eradication from the face of the planet has paved the way for my own self-interested ramblings. Passenger pigeons and Tazzy tigers, I owe you big time. Mammoths and dinosaurs, ditto. If you do ever come back, I promise

---

[*] 'Eruption' is, I believe, the correct collective noun for schoolchildren. There's also a 'tantrum' of toddlers and 'strop' of teenagers.

to buy you some beech mast/rodents/buttercups/fresh meat ...

Jim Martin, my editor at Bloomsbury, you are not extinct but thank you anyway. Thank you for buying me lunch that sunny summer's day and convincing me that this was a book worth writing. Thank you for your unwavering support and upbeat attitude throughout the entire process. It's been nothing less than a pleasure and a privilege from start to finish. You, Anna MacDiarmid and the guys at Bloomsbury are the best.

Matt Dawson (www.matt-dawson.co.uk), my insanely talented also-not-extinct illustrator, somehow you have managed to capture the essence of *Bring Back the King* with just a few, well-placed pencil strokes. I have no idea how you do it, but you are nothing short of brilliant. Thank you. No one draws a 'chirpy' dodo or a 'soulful' thylacine quite like you do. Your drawings make me happy. I like that. A lot.

Scientists, I salute you. In the course of writing this book, I have visited, Skyped, emailed, telephoned, texted, sent carrier pigeons to, stalked and generally bothered a veritable theory of researchers and other helpful people. Your insight, comments and kind words have been gratefully received.* Thank you Alberto Fernández-Arias, George Poinar, Roger Avery, Michael Benton, Ben Collen, Ross Barnett, Ryan Phelan, Stewart Brand, Darren Naish, John Hutchinson, Mary Schweitzer, Mike Buckley, Phil Manning, Sergio Bertazzo, Susie Maidment, Jack Horner, Arkhat Abzhanov, Semyon Grigoriev, Sergey Zimov, Michael McGrew, Akira Iritani, Yoshimi Kanzaki, Beth Shapiro, Adam Wolf, Insung Hwang, Roy Weber, Kevin Campbell, Arthur Caplan, Julian Hume, Jacquelyn Gill, Jing Liu, Mark Avery, Chris Stringer, Andrew French, Michael Mahony, Michael Tyler, Jan Stejskal, Koby

*There is no agreed collective noun for researchers, but reasonable suggestions include 'theory', 'whiteout' and 'bunsen'. Thanks to Matthew @MetaFatigue.

Barhad, Gareth Hempson, Richard Freeman, John Gurdon, Bob Meyer, Joni Mabe, Gil McVean, Mark Evans, Robert Plomin, Ben Johnson, Neil Hall, Dan Goldowitz, Mark Pallen, Gerd Kempermann, Chris Smith, Ben Minteer, Rhiannon Lloyd, Samantha Wisely, Bas Wintermans, John Zichy-Woinarski, Paul Meek, Rob Etches, Sayed-Morteza Hosseini, Mohammad H. Nasr-Esfahani, Stephen Seager, Robert Hermes, Michael Bruford, Andrew Torrance, Martha Gomez, Sam Turvey, Phil Seddon, Jeanne Loring, Anucha Sathanawongs, Richard Vigne, Robin Lovell-Badge, Susie Ellis, Joshua Akey, Oliver Ryder, Robert Etches, Hugh McLachlan, Gary Roemer, Matthew Hayward, Mark Jobling, Mark Witton, Steve Brusatte, Love Dalen, Ian Barnes, Tom Gilbert, William Holt, Tony Gill and Hugh Hunt for your time, your enthusiasm, your patience and your papers.

In addition, gold stars and the promise of a pint go to Ben Novak, Thomas Hildebrandt, Jack Price, Michael Archer, George Church, Malgosia Nowak-Kemp, Malcom Logan, Jeff Craig, Thomas Wynn, Michael Kjelland and Timandra Harkness for going the extra mile and taking the time to proofread and comment on the various drafts that I produced. At the risk of sounding like a luvvie at an awards ceremony, I couldn't have done it without you.

To my friends, thank you for putting up with me. When I said, 'I just interviewed a guy who specialises in the epigenetic control of embryonic limb development', you heard, 'blah, blah, blah.' And when I said, 'I've a feeling that induced pluripotent stem cells will become a stalwart of future high-tech conservation strategies', you heard, 'blah, blah, wart, blah.' But it didn't seem to matter. The room just went quiet for a little bit then the conversation started up again. Thank you Toni and Jon Harrington, Allison Botha, Justine Mallard, Eve, Jeremy and Harry, Caroline Forman, Sarah 'Stobbsy' Stobbs, Tracey Mafe, Andrea Warrener, Jane Bennett, Jess Semple, Isobel Collins, Sue Smith, Rachel Waters and Jo Brodie. Thank you for the blurry, booze-fuelled evenings, the friendship and the support. Thank you Brian,

Clare and Milly for the doggy day care. Phil Davidson and Meirel Whaites, thanks for being so lovely to my kids and for inspiring us all to find joy in Jurassic rock. Sara Abdulla, you got me into all this. Thank you for believing in me. Aĉiū, Paulius Tricys. Aš pasiilgau tavęs. Thanks also to Twinings, who make the best Earl Grey tea bags. I can report that I have drunk no fewer than 2,160 mugs of tea (milk, no sugar) during the writing process. Beat that if you can, Harkness.

To anyone else that I may have forgotten: apologies. You were amazing. I will make you a cuppa next time we meet. I might even throw in a custard cream.

And finally … To my husband, Joe. To my children, Amy, Jess and Sam. To my mum, Nijole, and to my very own hound dog, Higgs. Thank you for being there, always. Thank you for the warmth, the love and the laughter. Thank you for the frequent distractions, the endless cups of tea and for putting up with me not being there all the evenings and weekends that I spent working in my study. You were always on my mind.

Index

Abzhanov, Arkhat 60–61
Ailuropoda melanoleuca 258
Akey, Joshua 91
alligators 60
amber 21–22
Anning, Mary 43–44
Anoplolepsis gracilis 226
ant, yellow crazy 226
Archaeopteryx 58, 61
Archer, Michael 153, 156–60, 166–67, 171, 173, 178
artificial insemination 257–62
Asara, John 49–50
Ashmole, Elias 131
atavisms 59
Audubon, James 150
auk, great 16, 28
aurochs 16, 17
Avery, Mark *A Message from Martha* 149
Avery, Roger 22
avian malaria 245–46

back-breeding 212–13
banteng 214
Barnett, Ross 27
Batrachochytrium dendrobatidis 166
bats 245
bear, cave 27, 66–67
  polar 258
beaver 230–31
Ben-Nun, Inbar Friedrich 276
Benjamin 151–52, 155
Benton, Michael 25, 44
Bertazzo, Sergio 51
biodiversity 229–30, 242, 249–50, 266, 278–80
  islands 226–27
Biology of Reproduction 145
birds 58, 245–46
'Bob' 46–49, 51
Bolt, Usain 41
boobook, New Zealand 244
  Norfolk Island 244
Bos gaurus 214
  javanicus 214
  primigenius 16
'Brachy' 51
Brachylophosaurus canadensis 51
Brand, Stewart 30, 79, 139–40, 241, 243
Briggs, Robert 168–69, 171
Brown, Louise 177, 247
bucardo 11–17, 25, 28, 120–21, 142, 211–14, 220–21, 223, 237, 239
Bullockornis planei 153
Bush, George W. 206
bustard, houbara 238

Caloenas nicobarica 134
Campbell, Kevin 114–15
Canis lupus 228

Caplan, Arthur 88
Capra pyrenaica pyrenaica 12
Carroll, Lewis *Alice's Adventures in Wonderland* 126, 136
Carter, Peter 159
Castanea dentata 245
Ceratotherium simum cottoni 253, 268
s. simum 268
Chang, Il-Kuk 145–46
Charpentier, Emmanuelle 117
chestnut, American 245
chickens 58–63
wd25 145–46
Chicxulub asteroid 33–4, 58
Chironectes minimus 153
Chlamydotis undulata 238
Church, George 77–79, 114–18, 139, 142, 173, 185, 213, 219–20, 237, 242, 246
chytrid fungus 166, 220, 234–36, 246
classification of species 217–20
Clinton, Bill 205
cloning 12–15, 211–12
dogs 110–11
Dolly 105–107
Esfahan mouflon 238–39, 248–49
ferrets 242–43
frogs 167–71
frogs, gastric-brooding 171–78
humans 185, 188–89, 194–95, 206–209
Neanderthals 87–90
never be the same 212–13
passenger pigeon 142–44
tears of a clone 214–15
woolly mammoth 102–105, 107–10, 111–14, 121–22
cockroaches 26
Colgan, Don 158–59
Collen, Ben 26
Collins, Francis 88
condor, California 222–23, 238, 240
giant 27
conservation 235–39
black-footed ferret 239–43, 258, 275
genetic rescue 244–46
Cooper, Alan 134
coprolites 74–77
Corpen, Chris 163
crab, Christmas Island red 226, 227
Craig, Jeff 203–204
crane, white-naped 257–58
Cretaceous Period 9, 34, 38, 49–50
Crichton, Michael *Jurassic Park* 20–21, 23
Crick, Francis 17
CRISPR 116–17, 142, 194, 207
CRISPR-Cas9 248
Culex quinquefasciatus 245–46
Cuvier, Georges 24

Darwin, Charles 43, 191
de-extinction 9–10, 15, 17, 23–24, 30–31, 211–12, 278–80

all de-extinct but nowhere to go 223–24
Audrey Hepburn with whiskers 239–43
bucardo 11–17, 25, 28, 120–21, 142, 211–14, 220–21, 223, 237, 239
Christmas Island rat 224–28
decision time 220–23
genetic rescue 244–46
gut feeling 215–17
most unusual sperm bank 234–39
never be the same 212–13
passenger pigeon 137–44
playing God? 247–50
reintroduced species 228–32
tears of a clone 214–15
thylacine 158–61
what's in a name? 217–20
Dead Famous DNA 189, 199
Dediu, Dan 82
dengue fever 245–46
Denisovans 70
Dermochelys coriacea 236
'Dima' 102–105
Dinornis giganteus 134, 221
dinosaurs 22–23, 25, 220
blood cells 45–47
collagen 49–51
hunt for dinosaur DNA 52–57
medullary bone 47–48
teeth 38–39
DNA 9, 17
1000 Genomes project 191–95, 200
cloning 214–15
dodo 134–36
genetti spaghetti 17–20
hunt for dinosaur DNA 52–57
Jurassic spark 20–23
mummies 70–71
Neanderthals 73–74
preventing contamination 72–74
warts 188–89
Dodgson, Charles 126
dodo 9, 17, 27, 125–26
as dead as 126–30
back to Blighty 130–37
dogs, bear 26
Dolly 14, 105–107, 142, 170, 184, 211, 247, 274
dolphin, Yangtze river 223–24
dormouse, giant 128
Doudna, Jennifer 117
dove, mourning 148
drill 277

ecosystem engineers 229–31
Ectopistes migratorius 16, 137
Edwards, Robert 177, 247
electro-ejaculation 257–58, 260, 269
elephant birds 128
elephants 39–40, 105, 213, 215, 220, 246, 261, 277
elephant eggs 118–21
woolly mammoths 114–18, 121–23
emus 58
epigenetics 98, 201–204
Equus ferus przewalski 257
quagga quagga 16

Etches, Robert 249
European Society of Human Reproduction and Embryology 251
evolutionary biology 60–64
Extinct DNA Study Group 22–23
extinction 24–25, 278–80
  dodo 129–31
  mass extinctions 25
  passenger pigeon 138–39
  rhino, northern white 264–69
  thylacine 153–55
  Yangtze giant softshell turtle 266–67
  Yangtze river dolphin 223–24

Fernández-Arias, Alberto 13–15
ferret, black-footed 239–43, 258, 275
Firestone, Karen 158–59
Fleay, David 155
Folch, José 12, 239
fossils 8–9, 24, 26–27
  coprolites 74–77
  fossil DNA 19
  from bone to stone 42–45
fox, Santa Catalina 237–38
French, Andrew 175–77
frog, African clawed 169
  Darwin's 163
frog, gastric-brooding 162–66, 171–78, 220, 221, 246, 279
  northern 165–66
  southern 162–65
frog, great barred 172–76
  northern leopard 168–69
  Titicaca water 220–21
frozen specimens 54, 107–108, 171, 242
  Frozen Zoo 275–78

Galápagos tortoises 29
Galli, Cesare 271
Galton, Francis 197
gastrulation 173–74, 176
gaur 241
Gecarcoidea natalis 226
genes 18, 199–200
  FOXP2 83
  Hox genes 59–60
  MC1r 118
  Pitx1 61
genetic rescue 244–46
genetically modified organisms (GMOs) 218–19
genomes 17–18
  1000 Genomes project 191–95, 200
  CRISPR 116–17, 142, 194, 207, 248
  human genome 189–91
  Neanderthal genome 77–82
Gilbert, Tom 54, 71, 224–25
Gill, Jacquelyn 100–101, 266
gorilla 220, 236
Gorilla beringei beringei 236
Goto, Kazufumi 103–107
Grigoriev, Semyon 97–98
Grus vipio 258
Guardian, The 50

Gurdon, John 169–72, 176
Gymnogyps californianus 222

habitat 223–24
haem 46, 113
haemoglobin 115, 117
Hall, Neil 207–208
Harmon, Bob 46
hen, heath 16
Hermes, Robert 272–73
Hess, Roberta 21–22
Higuchi, Russell 53
Hildebrandt, Thomas 120, 258–62, 264–65, 269–73, 277, 279
Hogg, John and Lucille 240
Holt, William 261
hominins 54, 69, 89
Homo 26–27
  floresiensis 128
  heidelbergensis 82
Horner, Jack 45–47, 57–63
  How to Build a Dinosaur 59
horse, Przewalski's 257
horses 54, 72, 102, 105
Howard, JoGayle 241
Human Genome Project 77, 88
Human Tissue Act 2004 (UK) 206
humans 26–27
  interbreeding with Neanderthals 90–94
  language development 82–87
  oldest human DNA 54
Hume, Julian 129, 135
Hunt, Hugh 124
Hutchinson, John 41
Hwang, Insung 111, 113
Hwang, Woo Suk 110
Hydrodamalis gigas 28
Hypnomys mahoensis 128

Ice Age 100
Ice Age 98–101, 217, 265–66
in vitro fertilisation (IVF) 103, 175, 177, 204, 235, 247, 257, 72–73
Independent 15
International Union for the Conservation of Nature (IUCN) 218, 219, 279
intracytoplasmic sperm injection (ICSI) 105
iPS cells 276–78
Iritani, Akira 105–109
iron 55–57

Jurassic Park 20–21, 23, 50, 57, 234

kangaroos 27, 153
Kempermann, Gerd 204–205
Kennedy, John F. 205
keystone species 228–29
King, Thomas 168–69, 171
Kjelland, Michael 143–44

L'Estrange, Hamon 130
Lalueza-Fox, Carles 80, 81
language development 82–87
Lazarus Project 16, 17, 161–67, 234–35
  almost tadpoles 171–78
  early days 167–71

Lazzari, Giovanna 271
leopard, Formosan
 clouded 29
 snow 277
leopards 258
Leopold, Aldo *A Sand County Almanac* 237
Liem, David 162–63
Lindahl, Tomas 53–54
Linnaeus, Carl 127
lion, marsupial 27
Lloyd, Rhiannon 214
Logan, Malcolm 61, 62
Lonesome George 29–30
Longrich, Nick 39
Loring, Jeanne 274–77
Lynch, Vincent 116

Macroscincus coctei 29
MAGE 79, 194
Mahony, Michael 165–67, 172, 174, 234–37
Mallison, Heinrich 41
Malyi Lyakhovski 96–98, 111–12
mammoth, woolly 16, 20, 22, 27, 95–98, 213, 215, 220, 237, 242, 277, 279
 edit an elephant 114–18
 elephant in the room 118–22
 hello Dolly 105–107
 load of old bull 107–10
 mutts to mammoths 110–14
 recent extinction 99
 woolly legend 98–102
 woolly thinking 122–24
 young dreams 102–105
Mammuthus creticus 128
 primigenius 16, 95, 98, 220
Mandrillus leucophaeus 277
Mare aux Songes, Mauritius 135–36
'Martha' 139
Martin, Paul 27
mass extinctions 25, 278–80
 dropping like flies 26–30
Masters, George 157
Maynard, Charles 245
McCarthy, Robert 85
McGrew, Michael 106
McVean, Gilean 193–94
megafauna 26–27, 29, 266
melanin 44, 46
microbiomes 213, 215–16
Mikhelson, Viktor 102–103, 106
Mixophyes fasciolatus 172
moa, giant 134, 221
mouflon, Esfahan 238–39, 248–49
Multiplex Automated Genome Engineering (MAGE) 79
mummies 53, 70–71

Naish, Dareen 40
NASA 23
Nasr-Esfahani, Mohammad 238
National Cancer Institute (NCI) 168
National Human Genome Research Institute 190
Nature 71, 106, 207, 245, 251
Neanderthals 65–67, 220

let's talk Neanderthal 82–87
make me a caveman 77–82
meet the family 67–70
no sex please, we're Neanderthals 90–94
perhaps we could, but should we? 87–90
terrible smell 70–74
top of the plops 74–77
Neofelis nebulosa brachyura 29
New York Times 38, 77
Newell, Norman D. 157
Newton, Sir Isaac 191
Ninox novaeseelandiae 244
n. ndulata 244
Nothrotheriops shastensis 74
Novak, Ben 140–42, 146, 148–49, 213, 220, 248
Nowak-Kemp, Malgosia 132–34

Oncorhynchus mykiss 246
opossum, water 153
Osborn, Henry Fairfield 38
otters, bear 26
Ovibos moschatus 118
Ovis orientalis isphahanica 238
ox, musk 118

Pääbo, Svante 70–77, 83, 90–91
Paddle, Robert *The Last Tasmanian Tiger* 154–55
panda, giant 18, 219–20, 258, 261
panther, Florida 244
Panthera uncia 277
parthenogenesis 248
Parton, Dolly 106
Patagioenas fasciata 141–42
penguin, Magellanic 258
Pevzner, Pavel 50
Phelan, Ryan 30, 139–40, 241–42
pigeon, band-tailed 141–42, 146, 213
Nicobar 134, 137
pigeon, passenger 16, 17, 213, 220, 242, 248, 279
avian eclipse 137–44
don't cry, Daddy 144–50
pigeons 128–9
Pinguinis impennis 16
Pithovirus sibericum 216
Pleistocene 27, 67, 98–99, 115, 123
Pleistocene Park, Siberia 101–102, 123–24, 219
Plomin, Robert 200, 205
Poinar, George 21–23
Poinar, Hendrik 74–75
Powell, William 245
Presley, Elvis 10, 56–57, 105, 182–84, 220
Americans for cloning Elvis 184–91
burning love 179–82
essence d'Elvis 191–95
identical but different 201–204
infinite number of Elvises 204–206
jailhouse rock 206–209
seeing double 195–201
Pristis zijsron 221
Puma concolor coryi 244

quagga 16, 53, 212–13

Rafetus swinhoei 266
Rana pipiens 168, 169
Raphus cucullatus 27, 128
rat, black 224–25
  Christmas Island 224–28
Rattus macleari 224
Revive and Restore 140, 148–49, 241–43
Rheobatrachus silus 16, 162
  vitellinus 165
rhino, black 255, 268
rhino, greater one-horned 255
rhino, Javan 255
rhino, northern white 253–56, 279–80
  artificial insemination 257–62
  dead rhino walking 264–69
  last chance to survive 262–64
  rhino eggs 269–73
  rhino progger 252–53, 256
  Y factor 273–78
rhino, Sumatran 255
rhino, white 255
rhino, woolly 27, 107
Rhinoceros unicornis 255
Rhinoderma darwinii 163
Richard III 206
Riversleigh, Queensland 153, 157
Roberts, Greg 163
Roemer, Gary 244, 246
Rolling Stone 183
Ryder, Oliver 274–75

sabre-toothed cats 26, 66
Sacrison, Stan and Steve 36
Saffery, Richard 203
Saguinus bicolor 221
Sarcophilus harrisii 159, 245
sawfish, chainsaw-nosed narrowsnouth 221
Schaaffhausen, Hermann 80
Schweitzer, Mary 45–51, 53, 55–57
Science 26, 49, 164
sea cow, Steller's 28, 222
sea cucumber, pineapple 221
Seager, Stephen 256–58, 262
Sesame Street 100
Shanidar Cave, Iraq 65–66
Shapiro, Beth 134–36, 139–41, 149
skink, Cape Verde giant 29
sloth, giant 27, 74
Solecki, Ralph 66–67
sperm banks 234–39
Sphenicus magellanicus 258
Spielberg, Steven 20
'Stan' 33–4, 36
Stejskal, Jan 255, 276
Steller, Georg Wilhelm 28
stem cells 110, 170, 273–76, 279
Steptoe, Patrick 177, 247
sylvatic plague 241, 243

Tabin, Clifford 61
tamarin, pied 221
Tasmanian devil 159, 245
TaurOs Project 16–17
TEDx 30, 153, 173, 242
Telmatobius culeus 220

Thelenota ananas 221
thylacine 151–56, 220
  in a pickle 156–61
Thylacinus
    cynocephalus 153
tiger 220, 238, 261
  Tasmanian *see* thylacine
Time 106, 173
tinamou 221
Titchmarsh, Alan 100
Tkach, John 22
Torrance, Andrew 219
Triceratops 17, 38–39
trout, rainbow 246
turtle, leatherback 236
  Yangtze giant
    softshell 266–67
Turvey, Samuel 224
twins 195–99, 201–205
Tyler, Mike 163–65, 167,
    171
Tympanuchus cupido cupido 16
Tyrannosaurus rex 33–4
  acid test 47–52
  burning love 35–7
  from bone to stone 42–45
  hunt for dinosaur
    DNA 52–57
  impossible dream 57–60
  'little round red things'
    45–47
  shape shifters 60–64
  what could possible go
    wrong? 37–42

Urban, Mark 29
Urocyon littoralis 238
Ussher, James 24

Velociraptor 58
Vigne, Richard 269
viruses 216–17

Wakayama,
    Teruhiko 107–108
Walter, Lawrence 'Lawnchair
    Larry' Richard 233–34,
    249
'Wankel Rex' 36, 45
Wankel, Kathy 36
Watson, James 17
Weber, Roy 112–13
Werdelin, Lars 26–27
Westbrook,
    Roberta 155–56
white-nose syndrome 245
Wilmut, Ian 247, 274
Wilson, Allan 53
Wintermans, Bas 216
Wisely, Samantha 240–41
Wittmeyer, Jennifer 48
wolf, grey 228–29
  Tasmanian *see* thylacine
wombat, giant 27
World Wildlife Fund 237
Wynn, Thomas *How to Think
    Like a Neanderthal* 81,
    84, 86

Xenopus laevis 162, 169

Yamanaka, Shinya 273–74
Yellowstone National Park,
    USA 228–29

Zimov, Sergey 101–102,
    123